FINANCIAL PERFORMANCE OF AGRO-BASED INDUSTRIES

THE CASE OF SUGAR INDUSTRY IN KARNATAKA

FINANCIAL PERFORMANCE OF AGRO-BASED INDUSTRIES

THE CASE OF SUGAR INDUSTRY IN KARNATAKA

T.S. DEVARAJA

ANMOL PUBLICATIONS PVT. LTD
NEW DELHI - 110 002 (INDIA)

ANMOL PUBLICATIONS PVT. LTD.
H.O.: 4374/4B, Ansari Road, Daryaganj,
New Delhi-110 002 (India)
Ph.: 23278000, 23261597
B.O.: No. 1015, Ist Main Road, BSK IIIrd Stage
IIIrd Phase, IIIrd Block,
Bangalore - 560 085 (India)
Visit us at: www.anmolpublications.com

Financial Performance of Agro-Based Industries: The Case of Sugar Industry in Karnataka

First Published, 2009

ISBN 978-81-261-4047-3

ICSSR Consultant: Prof. R.K. Mishra

[This book is published with financial support from the Indian Council of Social Science Research, New Delhi. However, the responsibility for the facts stated and conclusions reached is entirely that of the Author and ICSSR, New Delhi bears no responsibility for them.]

PRINTED IN INDIA

Printed at Mehra Offset Press, Delhi.

CONTENTS

ACKNOWLEDGEMENT

I wish to take as a foremost duty to express my sincere and heartfelt gratitude to the Chairman, Member Secretary, Director (Research Project Division), and the team of ICSSR for sanctioning the project and providing the financial support for smooth execution of the project. My respectful thanks to all the officials of the Research Survey and Publication Division of ICSSR especially Dr. Prem Prakesh Pandey for his moral support and considering this book for publication grant.

I sincerely thank all the Managing Directors, Accounts and Finance Officers of the sample sugar factories operating in Karnataka for their valuable and timely supply of required data for the research. My respectful thanks are due to the Managing Director and Manager of Karnataka Co-operative Sugar Factories Federation, Bangalore; the Secretary, South India Sugar Mills Association (Karnataka Branch), Bangalore; the Secretary, Karnataka Brewers and Distillers Association, Bangalore, and the Commissioner of Cane Development and Distilleries of Sugar, Government of Karnataka, Bangalore, for their valuable guidance and for providing required data to carry the research work.

I express my sincere gratitude and thanks to the Secretary, Indian Sugar Mills Association, New Delhi; the Managing Director, National Federation of Co-operative Sugar Factories, New Delhi; the Secretary, All India Distilleries Association, New Delhi; the Secretary, Sugar Technologists Association of India, New Delhi; and the General Manager, Indian Sugar

EXIM Corporation Ltd., New Delhi; the Chief Director, Directorate of Sugar, Ministry of Consumer Affairs, New Delhi for their kind response to the scheduled questionnaire to carry out this work.

I sincerely thank The Anmol Publications Pvt.Ltd. New Delhi, for publishing this work with greatest care.

My very sincere gratitude is acknowledged to Mr. G.Nijaguna, Assistant Professor Mr.Nageshwar Rao and Mr. Manjunath, Research Assistants for their assistance in collation of statistical data and analysis.

I am grateful for The Honorable Vice-Chancellor, University of Mysore Prof. Dr. V.G. Thalwar, who is always encouraging and supportive of my academic and research works.

My very sincere gratitude to The Director and Professor Vasudev, who is most kind and supportive of the academic works at our University Post Graduate Centre, Hassan, University of Mysore.

My words of indebtedness remain for Professor B.H. Suresh, under whose research supervision of my doctoral thesis, I strengthened my academic career.

I would like to thank the Department of Commerce, especially Prof. Yashwanth Dongre, with whose suggestions I was able to reenvision the entire comprehensiveness of my work.

My especial acknowledgement of thanks goes to my friend Ravindra Babu, Lecturer in English at our University Centre who always encouragingly shared my academic interests.

I thank my family for the greatest understanding and support they showed while I worked on to complete this.

Date: 15-12-2008 **T.S.DEVARAJA**

PREFACE

My justification for bringing out this research report in the form of book on an already widely discussed subject is the conviction that it still affords room for analytical inquiry. I presume I have tried to explore aspects of the financial performance of sugar industry in Karnataka which have either not been perceived at all or have been but dimly seen.

The available sources of literature on financial performance of sugar industry are highly technical in character and so diverse that it becomes difficult to locate required information in one place. Moreover, most of the financial literature is replete with statistical formulae with explanations in financial terms. It has been provided with up-to-date information in a comprehensive way, dealing with all practical aspects and described as briefly as possible with financial data.

A genuine effort has been made to explain the present status of Financial Performance of Agro-Based Industries: The case of sugar industry in Karnataka. The present study has been divided into eight chapters namely; Introduction about the agro-based industries and sugar industry scenario of India and Karnataka state; Overview of Sugar Industry in India; Status of sugar industry in Karnataka; Financial performance of sugar industry in Karnataka: concepts and context; Analysis of Financial performance of sugar industry in Karnataka; Comparative analysis of financial performance of sugar industry in Karnataka; Policy Recommendations and suggestions and finally Conclusions. The results of this research on Financial Performance of Agro-Based Industries: The Case of Sugar Industry in Karnataka are presented in seven sub-heads; *viz.*, pattern of ownership and performance of sugar industry in Karnataka; performance and growth of sugar factories; financial analysis of sugar factories; statistical analysis of the sugar factories; capacity utilization of sugar factories; cost of processing sugarcane into sugar and overview of the functioning of the sugar factories through SWOP (T) management technique. The important financial tools used in the present work are Ratio Analysis, Dupont Analysis, Breakeven Analysis and Safety Margin to evaluate the financial strength of selected sugar factories in Karnataka. Statistical tools like Standard Error, *t*-Test and Compound Growth Rate are also used to test the validity of data used for the present research study.

Date: 01-12-2008

Place: **Mysore** **T.S.DEVARAJA**

"Great discoveries and improvements invariably involve the cooperation of many minds. I may be given credit for having blazed the trail, but when I look at the subsequent developments I feel the credit is due to others rather than myself"

— Alexander Graham Bell

"Every kind of peaceful cooperation among men is primarily based on mutual trust and only secondarily on institutions such as courts of justice and police"

—Albert Einstein

"Coming together is a beginning. Keeping together is progress. Working together is success"

— Henry Ford

"The battle of competition is fought by the cheapening of commodities."

— Karl Marx

"Leadership is based on inspiration, not domination; on cooperation, not intimidation"

— William Arthur Ward

LIST OF TABLES

LIST OF GRAPHS

LIST OF APPENDICES

ABBREVIATIONS USED IN THE STUDY

AICPR	All India Coordinated Research Project
AVS	Average Variable Cost
ATC	Average Total Cost
AVG	Average
BEP	Break Even Point
CU	Capacity Utilisation
CP	Cost of Processing
CGR	Compound Growth Rates
CS	Co-operative Sector
DER	Debt-Equity Ratio
DMS	Diffuser Mechanism System
DGCI & S	Directorate General of Commercial Intelligence & Statistics
DSFA	Deccan Sugar Factories Association
EM	Equity Multiplier
E10	Ethanol 10
EU	European Union
FI	Financial Indicators
GOI	Government of India
GATT	General Agreement for Trade and Tariffs
Ha.	Hectare

ICRA	Investment Information and Credit Rating Agency of India Limited
ISEC	Indian Sugar Export and Import Corporation Ltd.,
ISO	International Standard Organisation
ISMA	Indian Sugar Mills Association
KSFCSF	The Karnataka State Federation of Co-op Sugar Factories Ltd.,
KBDA	Karnataka Brewers & Distillers Association.
KSCMF	Kerala State Co-operative Marketing Federation
KSI	Karnataka Sugar Institute
KL	Kilo Litre
MT	Metric Tones
MM	Mile Meters
MW	Mega Watts
MARKFED	Marketing Federation
NFCSF	The National Federation of Co-op Sugar Factories Limited
NSI	National Sugar Institute
NI	Net Income
NCDC	National Co-operative Development Council
PI	Physical Indicators
PACS	Primary Agricultural Co-operative Society
ROIC	Return on Invested Capital
PM	Profit Margin
ROI	Return On Investment
ROE	Return On Equity
R & D	Research and Development
RRB	Regional Rural Banks
SMP	Statutory Minimum Price
SAP	State Advised Price
SISMA	South Indian Sugar Mills Association

SISSTA	The South Indian Sugarcane and Sugar Technologists Association
SM	Safety Margin
SE	Standard Error
SDF	Sugar Development Fund
SWO (P) T	Strength, Weakness, Opportunities (Problems) Threats
SF	Sugar Factories
TE	Total Equity
TFC	Total Fixed Cost
TR	Total Revenue
TC	Total Cost
TVC	Total Variable Cost
TSP	Total Sugar Production
TCD	Tones Crushing capacity per Day
TAT	Total Assets Turnover
TA	Total Assets
TC	Tariff Commission
VAT	Value Added Tax
WRT	With Refers To
WTO	World Trade Organisation

Chapter 1

INTRODUCTION

The Indian sugar industry is a key driver of rural development, supporting India's economic growth. The industry is inherently inclusive supporting over 50 million farmers and their families, along with workers and entrepreneurs of almost 550 sugar mills, apart from a host of wholesalers and distributors spread across the country. Indian sugar industry is a cyclical industry, as on one hand it serves the domestic market, the largest in the world. Sugar is a sector of significant importance to the national economy. While consumption has been growing historically, the production has been cyclical. At present, the sugar industry is regulated across the value chain. Investments in by-products are at a nascent stage, and the sector has struggled to generate a return on invested capital in excess of its cost of capital in most of the years, primarily due to a high mandated fixed cane price and a volatile sugar price. Emerging businesses like fuel ethanol and structural changes globally including the removal of European Union subsidies have provided new horizons for the sector. The sector today has transformational opportunities that would enable it not only to continue to serve the domestic markets but also emerge as a significant carbon credit and power producer and support an ethanol-blending programme of E10 and beyond. However the sugar regulations would have to evolve to facilitate the transformation of the sector.

In an era where there is a need for inclusive growth, the sugar industry is amongst the few industries that have successfully contributed to the rural economy. It has done so by commercially utilising the rural resources to meet the large domestic demand for sugar and by generating surplus energy to meet the increasing energy needs of India. In addition to this, the industry has become the mainstay of the alcohol industry. The sector also has a significant standing in the global sugar space. The Indian domestic sugar market is one of the largest markets in the world, in terms of volume. India is also the second largest sugar producing nation and remains a key growth driver for world sugar, growing above the Asian and world consumption growth average.

However, fundamental changes in the consumer profile and the demonstrated ability of the sector to continuously ensure availability of sugar for domestic consumption has diluted the need for sugar to be considered as an essential commodity. According to a recently conducted nation wide survey, nearly industrial, small business and high-income household segments consume 75 per cent of the total non-levy sugar. Further, even for a low-income household, 10 per cent increase in sugar price would result in less than one per cent increase in the three monthly food expenses. While the sector grows in stature and continues to play a key role in the economy, it is expected to face some significant challenges. There is lack of alignment between sugarcane and sugar prices. As a result, it leads to cane payment arrears and induces cyclicality. The arrears typically result in the eventual need for government support packages, while the pronounced cyclicality destabilises the sector revenues. The average sugarcane yields have also, at best, stagnated, and the average recovery is amongst the lowest in comparison with key sugar producing nations. Large sugar inventory exposure and sugar price volatility also results in high sugar price risk for the sector. In the past ten years, on an average basis, even the large listed sugar firms have struggled to generate return on Invested Capital over and above their cost of capital. This is primarily due to high mandated fixed cane prices and volatile sugar prices.

In 2007, the domestic sugar consumption is estimated to be 19.5 million metric tonnes. It is expected that the drivers for consumption i.e. the GDP growth and population growth would continue to grow at current rates. Based on the past ten year's growth in consumption and estimates from various independent sources, it is expected that in 2017, the domestic sugar consumption would be approximately 28.5 million metric tonnes. Given the high cost of imports and the strategic importance of food security, India would need to target its production in excess of domestic consumption. Given the past trend in production cyclicality, sugar equivalent to 1.5 times of consumption i.e. an additional 3.5 million MT of sugar would need to be produced by 2017.

Agro-Based Industries Scenario in India

Agro-processing industries have a prominent place in the manufacturing sector in India. They make a significant contribution directly as well as indirectly by stimulating development in the agriculture and rural sectors. The scope of the agro-processing industry encompasses all operations from the stage of harvest till the material reaches the end users in the desired form, packaging, quantity, quality and price. Agro-processing is now regarded as the sunrise sector of the Indian economy in view of its large potential for growth and likely socio-economic impact specifically on employment and income generation. Some estimates suggest that in developed countries, up to 14 per cent of the total work force is engaged in agro-processing sector directly or indirectly. However, in India, only about 3 per cent of the work force finds employment in this sector revealing its underdeveloped state and vast untapped potential for employment. Properly developed, agro-processing sector can make India a major player at the global level for marketing and supply of processed food, feed and a wide range of other plant and animal products.

The important agro-processing industries in India classified on crop and commodity wise are; Rice Processing Industry; Wheat Processing Industry; Processing of Coarse Cereals; Oilseeds Processing; Processing of Pulses; Processing of Fruits and Vegetables; Cotton Processing Technology; Sugarcane Processing Industry; Processing of Jute; Processing of Animal Produce; Fish and Fish Products Processing; Processing of Plantation Crops; Processing of Commercial Crops; Processing of Medicinal and Aromatic Products; Apiculture Produce Processing; Floriculture; and Processing of Traditional Foods Research and Development in Agro – processing Industry.

At the start of the twentieth century, Indian agriculture was in a stage of subsistence. But by the end of the year 2007-2008, India started producing about 800 million tonnes of biological materials per year including food grains, oilseeds, fruits, vegetables, sugarcane, milk, eggs, meat, fish, tea, coffee, fiber crops, floricultural produce, forest produce and so on. However, their market potential is not being fully realised due to poor post harvest management and inadequate infrastructure and programme for processing of agro-produce.

Growth of Agro-Processing Sector in India

Starting with a small number of processing facilities in 1950-51, a fairly well spread network of processing facilities has developed in the Country. Various estimates suggest the number of processing units in 2007-2008 as: atta chakkis and small hammer mills - 2,70,000, rice hullers - 90,000, rice shellers - 11,000, huller-cumshellers - 12,000, modern rice mills - 30,000, bullock and electricity operated oil ghannis - 2,00,000, oil expellers - 55,000, dhal mills - 12,000, roller flour mills - 700, rice flaking and puffing units - 2,000, bakery units - 54,000, solvent extraction plants - 700, vanaspati plants - 100, fruits and vegetables processing plants - 5,000, dairy plants - 450, cold storage units - 3,000, licensed units in organised sector for meat processing - 165, pork processing units – 144, fish processing units - 18 and so on.

Major problems faced by these agro-processing units have been:(a) low capacity utilisation, (b) poor recovery of the finished product from the raw materials, (c) problems of arranging adequate working capital and its management, (d) low product quality and (e) unreliable assured power supply. Strong research and development support will have to be continued to over come these and many other problems to ensure that our agro-processing technology becomes competitive at the global level. As stated earlier, in spite of the problems, agro-processing technology in India has continued to make steady progress towards modernisation.

The important Research and Development activities carried out in agro-processing in India during the last fifty years categorised as follows:

(a) Studies on physical, biochemical, nutritional, and engineering properties and characteristics of different food, feed, fibre, and industrial raw materials.

(b) Response studies of different biological materials w.r.t. their storage, handling, and moisture conditioning.

(c) Refinement of traditional equipment and processes for production of different foods, feeds, fibres and fuel materials for better quality, higher capacity, energy efficiency, and reduced drudgery to workers.

(d) Development of new produces and processes for better nutrition, convenience and taste.

(e) Enhancement of shelf life of the produces, safe storage and packaging and development of better performing materials.

(f) Better economic utilisation of agricultural residues, by-products and recycling of wastes.

(g) Design and Development of instruments and equipment for post harvest operations and their evaluation, feasibility analyses, field trails/ multiplications valuation etc.

(e) Design, layout planning and development of pilot plants, agricultural produce bulk handling systems and area specific agro-processing models.

(f) Studies and modeling/simulation of post harvest systems and industry for the purpose of optimisation, forecasting and policy analyses.

(g) Energy auditing and use of non-renewable sources of energy for post harvest operations.

(h) Product quality analyses, sensory evaluation and consumer acceptance studies.

(i) Work conditions, safety and pollution control.

SWOT Analyses of Agro-Processing Industry Infrastructure in India

Strengths

(a) Round the year availability of raw materials.

(b) Social acceptability of agro-processing as important area and support from the central government.

(c) Vast network of manufacturing facilities all over the country.

(d) Vast domestic market.

Weaknesses

(a) High requirement of working capital

(b) Low availability of new reliable and better accuracy instruments and equipments

(c) Inadequate automation w.r.t. information management.

(d) Remuneration less attractive for talent in comparison to contemporary disciplines.

(e) Inadequately developed linkages between R&D labs and industry.

Opportunities

(a) Large crop and material base in the country due to agro-ecological variability offers vast potential for Agro-processing activities.

(b) Integration of developments in contemporary technologies such as electronics, material science, computer, bio-technology etc. offer vast scope for rapid improvement and progress.

(c) Opening of global markets may lead to export of our developed technologies and facilitate generation of additional income and employment opportunities.

Threats

(a) Competition from global players

(b) Loss of trained manpower to other industries and other professions due to better working conditions prevailing there may lead to further shortage of manpower.

(c) Rapid developments in contemporary and requirements of the industry may lead to fast obsolescence.

Export Trends and Opportunities

India has been a traditional exporter of raw agricultural products like spices. Export of raw products has resulted in huge loss to Indian economy. After GATT agreement and WTO membership, processed products manufactured as per international norms only offered at competitive prices, can be exported. However, our processed products mostly do not meet the international standards. India's share in over US$ 300 billion world trade in agricultural commodities is less than 1 per cent Agricultural exports used to be of the order of 30.6 per cent of the total exports during 1980-81, which came down to 19.4 per cent by 1990-91. Currently, it is at about 16 per cent due to rapid growth in other sectors as well. Processed fruit and vegetable products have considerable export potentials and if it is properly utilised, growers, processors, traders as well as national economy will benefit. It requires correct assessment of world market, high quality of raw produce, high quality of processed product and competitive production cost.

Plans and Strategy

The objectives of agro-processing programmes in India should be to:

(a) Minimize product losses,
(b) Add maximum value,
(c) Achieve high quality standards,
(d) Keep processing cost low, and
(e) Ensure that a fair share of added value goes to the producer

To achieve these objectives following strategies are suggested:

(a) National plan for improvement and extension of agro-processing technology at farm, traditional small industry and modern industry levels should be prepared. The plan should take into account the diversity in resources and needs of different regions in the Country. It should include programme details and implementation schedule for the first four or five years. The progress of plan implementation should be periodically reviewed to allow adjustments and corrective measures, and to develop programme details for the years beyond the period under review.

(b) Thrust areas for research and development should be identified and medium term research and development programme should be prepared and implemented to support the national plan for improvement and extension of agro-processing technology at different levels. Treatment and utilisation of effluents from agro-processing industry should be included in the Research and Development programme.

(c) Emphasis should be put on the establishment of new agro-industrial plants in the production catchments to minimize transport cost, make use lower cost land and more abundant water supply, create employment opportunity in the rural sector and utilize process waste and by-products for feed, irrigation and manure.

(d) Infrastructure in the production catchments selected for agro-industrial development should be improved. Because of uncertain grid power supply to rural areas, decentralised power generation using locally available resources may become an integral part of agro-industrial development. Similarly, if the raw materials and processed products are perishable or semi perishable in nature, cold chain will have to be established.

(e) The national plan should provide for management of agro-industrial activities in the catchments area, both by private companies and individuals as well as co-operatives.

(f) Financial incentives and support should be provided on liberal scale to promote the modernisation of agro-processing industry and for establishing new such industries in production catchments.

(g) Arrangements to supply market information to the farmer and agro-processor should be put in place.

The review of literature in the study attempts to take stock of research works carried out by various scholars in the areas closely related to the topic. It is expected that such an attempt will enable the researchers in planning the proposed study on scientific lines with due consideration to appropriate the critical framework. There is large number of studies conducted on economic efficiency of sugar industry in various regions of the country and abroad. The available literature on the subject has been reviewed and presented in four sub-heads, keeping in view the objectives of the study.

1.1 Performances of sugar industries

1.2 Ratio analyses

1.3 Compound growth rates

1.4 Agri-related institutions and agro-based industries.

1.1 Studies on the Performance of Sugar Industry

Gupta *et al.* made a detailed analyses of consumption of sugar in India using regression model and reported that the high cost of production is responsible for the low per capita consumption of sugar. They introduced time component and got the regression equation

$$X = 5.3599+0.00025Y-0.0484t$$

Where X = Per capita consumption

Y = Selling price per quintal, T= Time factor

They worked out elasticity of demand at 0.017 and concluded that the sugar consumption is income-inelastic. Ostensibly this study has no direct relevance to the present investigation. But what is implicit in their study was the need for reduction in the cost of production to increase the consumption level. In fact, they did make an attempt in their study to show how profitability varied with changes in free sales price, which had a direct bearing on profits of the industry.

(1971)

Hirunandez et. al. in their study relating to sugarcane processing industry have opined that economists have used simple tools i.e. arithmetic average for estimating economies of scale on the basis of empirical data but such an attempt gives rough estimations about the most optimum capacity size group in respective industry. It is however difficult to conclude exactly as to what particular point the profits will be maximum. For this, it is necessary to estimate either profit function or cost functions from the available data. These functions enable us to determine the optimum capacity as the one having the highest profit or the lowest cost.

(1971, p.211)

Pillai estimated the cost of production of sugar for the year 1970-71 for four states namely, Tamil Nadu, Pondichery, Andhra Pradesh and Karnataka at Rs.142.60 per quintal, Rs.135.93 per quintal and Rs. 138.67 per quintal respectively.

He compared these cost with the ex-factory levy price fixed by the Government. He concluded that the Government did not provide adequate margins for depreciation, rehabilitation, increased price of gunny bags, extra cost of packing and finally increase in interest rate. This has caused heavy drain of profits of the sugar industry and the average loss for the four states according to him was to the extent of Rs. 15 per quintal. He also stated that the overall loss ranged from Rs. 4 lakhs to Rs.41 lakhs. Finally he concluded that there was urgent need for change in the price fixing formula especially in the south. The low levy price fixed and the resulting loss to the sugar factories, as found out by this study is an important consideration from the point of view of the present investigation.

(1971, pp.713-717)

Tariff Commission made a fresh enquiry into the cost structure of the sugar industry. It selected 59 sugar units out of 229 for investigation. For Karnataka, it assumed 146 days of crushing and 10.34 per cent recovery. The conversion cost estimated by the commission was Rs. 41.68. The breakup was as follows: Salaries and wages Rs. 10.31; power, fuel and stores Rs. 6.31; repairs and maintenance Rs. 3.91; overhead cost Rs.2.76; packing Rs. 4.36; selling expenses Rs. 0.18; depreciation Rs. 10.75 and cane transport Rs. 2.00. After making allowance for by-products, the total average conversion cost was estimated at Rs.41.40. It also concluded that increase in cane price and implementation of wage rates recommended by second wage board had significantly affected the cost. Another significant observation made by the commission was about the minimum cane price. It opined that the minimum cane price tends to become purely notional and had some relevance only for theoretically working out the ex-factory cost of sugar. It also felt that no new factory should be allowed to be established within the radius of 30 miles from an existing factory. Commission felt that a return of Rs.12.60 per quintal as reasonable to ensure a fair return on the invested capital. In calculating depreciation cost the commission followed written down value method in accordance with the Income Tax rules.

(1973)

Mohite stated that the rapid growth of sugar industry has been helpful to a great extent in stabilising sugar production and reducing chronic shortage of sugar in the country. The co-operative form of organisation has lent itself as a powerful instrument in the development and mobilisation of natural, human and financial resources of the state, towards sugar industry and has helped in reducing the backwardness of the country by improving the socio-economic conditions of rural areas in the state.

(1974)

The study of Parthasarathy also refers to sugar factories in Andhra Pradesh. It is an empirical study about the variations of cost structure of sugarcane in Andhra Pradesh. He arrived at certain findings with the help of sophisticated statistical tools and techniques with respect to cost of cultivation of sugarcane and fixation of cane price in Andhra Pradesh. However he stressed that, there is need for fixing reasonably fair and remunerative price for cane on the basis of sugar recovery and returns from alternative competing crops.

(1974, pp.188-189)

Sampath Committee made a detailed study of problems relating to the viability of new sugar factories. It estimated the establishment cost of sugar factory of 1250 tones capacity at Rs. 550 lakhs. The committee felt that there are still large pockets of sugarcane to the country, which had to be tapped for sufficient supply of sugarcane in the factories. The committee expressed that "viability of a factory would mean the unit should not only break-even after meeting the working expenses, interest on borrowings, depreciation on plant and machinery but should also be able to declare a reasonable dividend on equity capital". It classified Karnataka under high recovery area and the estimated conversion cost was Rs.50.00 per quintal. The average depreciation allowance made was at the rate of six per cent on the original cost of sets. Another significant suggestion made by it was the need for regular rebate on excise duty since giving a higher quota of free sale sugar to the new factories is not a workable measure in the long run. Because higher free sale quote would make the availability of the required minimum quantity for distribution through fair price shops difficult. It concluded that the incentive should be given on a long-term basis.

(1974)

Sugar Inquiry Commission also known as Bhargava Commission made a through inquiry of the problems of processing sugarcane into sugar, *gur* and *khandasari*. The committee held that extra realisation from free-sale sugar should be shared between the processor and the cane framers, at the ratio of 50:50. Some of the positive findings of the study were: (1) for meeting increasing cane requirements there was need to intensify cane development and also to bring more area under sugarcane. Therefore, it is essential to fix a remunerative price for sugarcane; (2) actual crushing period during the preceding year instead of 'stipulated optimum' should be considered to fix the minimum sugarcane price for the individual factories; (3) the scheme of linking the prices of sugarcane supplied by a cultivator to its quality cannot be enforced on all India basis because of practical difficulties; (4) incentives should be provided to persuade cane growers to enter into agreement since erratic supply of cane to factories can be greatly stabilised if cane growers could be persuaded to enter into agreements; (5) in fixing the minimum cane price, the commission felt that the statutory minimum cane price should be fixed so as not only to cover the cost of cultivation but also the return from cane was better from other alternative crops in which technological break-through had already been achieved. However, this should be varied in proportion to changes in the prices of other competing crops.

(1974)

Anand gave one of the special features of sugar industry as of recurring five yearly cycle, two years of rising trend, one year of peak production, followed by two years of decreasing production. Further he stressed the importance of sugarcane development programme for reducing intensity of this phenomenon. Apart from the problem of cyclical nature of production of sugar and variation in capital investment and sugar recovery, the sugar industry experiences in interstate difference in the cost of production per unit of sugar. The cost of production of sugar as estimated by Dhanuka (1975) in Bihar varied between Rs.107.05 and Rs.270.00 per quintal, the biggest and the lowest being in Uttar Pradesh and Bihar respectively.

(1975, pp.263-267)

According to Kamat, co-operative sugar factories have not only brought the economics of large-scale production to the members but also provided them with opportunities to organize and manage their own business. The members are trained in democratic and business ways.

(1975, p.175)

According to Kharkar now a days, there is an emphasis on increasing production capacity for the factory; underutilisation of capacity leads the factory to loss. The author has specifically pointed out that economics of sugar factory is disturbed largely because of low productivity and crushing of sugarcane of non-share holders who are not regular members and supply the cane of poor quality. This statement, however, seems to be particularly invalid for the reason that the recovery percentage of sugarcane to factory is quite encouraging.

(1975, pp.15-18)

Moni has made a study on prospects and problems of Tamil Nadu sugar industry. He found out that the average cost of production in Tamil Nadu and Pondicherry factories was Rs. 228.58 per quintal for the year 1974-75 as against government of India's calculation at Rs.211.30 per quintal. Out of this average cost of production of Rs.228.58 the conversion cost was estimated at Rs.66.44. Considering this actual average cost he reported that the levy price should have been fixed by the Government at Rs.179.51 per quintal. But the levy price was fixed at Rs.151.66 (the cost estimated by him was for an average recovery of 8.45 per cent and an average duration of 196 days). He concluded that the prospects for the Tamil Nadu sugar industries appeared rather gloomy on account of the policy of the Government on levy price fixation.

(1975, pp.17-20)

Ferron has rightly said that there is a necessity of determining the most economic size of the sugar mill in order to safeguard the industry from the problem of diseconomies. The basic consideration in determining the milling capacity of sugarcane mill must therefore, be illustrated as the basis of production cost and performance figures of factory.

(1976)

Mittal has given some of the suggestions for improving economies of sugar industry as follows:

1. Adoption of techniques to reduce losses of sugar.
2. Utilisation of by-products.
3. Increasing productivity of sugarcane.
4. Reduction in store consumption.
5. Changing formula of levy prices fixation and
6. Payments to cane growers according to cane quality rather than weight.

(1976, p. 281)

Mohite opined that the rapid growth of sugar industry has been helpful to a great extent in stabilising sugar production and in reducing chronic shortage of sugar in the country. The co-operative form of organisation has lent itself as a powerful instrument in the development and mobilisation of natural, human and financial resources of the state towards sugar industry and has helped in reducing the backwardness of the country by improving the socio-economic conditions of rural areas in the state.

(1974)

Phadnis while stressing the need for optimum capacity, he pointed out that as size of plant increases beyond a certain limit there exist diseconomies of scale due to inefficient management of the plant. If the factory is run at optimum capacity there will be efficient management which ultimately leads to reduction in the cost of production that maintains consistency of high price for cane growers.

(1976, p.95)

According to Thaper some of peculiar features of the sugar industry are as follows:

(a) The industry is seasonal.
(b) Duration of season varies from one factory to other and also from the year to year.
(c) Sugar content in cane and its recoveries also vary at random from one factory to another and
(d) Variation in capital investment.

(1976)

Karche has published an article on co-operative sugar industry in *Maratavada,* a lead industry for development and pointed out that the co-operative sugar factories not only increase the income of the members of the factory but also create large employment potential to the local people. The co-operative sugar factories with a crushing capacity of 1250 tonnes per day has got the employment potential of 400 permanent workers and 800 seasonal workers in the factory itself for the harvesting and for transporting of sugarcane; 6000 male and female, 100 trucks and 800 bullock carts are employed for period of nearly six months.

(1977, pp.15-20)

Kawade determined optimum use level of production capacity of the co-operative sugar factories in Maharastra on the basis of the behaviour of operating and total costs relative to installed capacity of the sugar factories. Two cost functions have been estimated on the basis of the data on average total cost (ATC) and average variable cost (AVC) per quintal of sugar in relation to installed capacity of the sugar factories (T) average total cost and average variable cost were regressed separately on the installed capacity and the following empirical relation of the quadratic type were obtained.

(1) $ATC = 246.24 - 0.010053\ T + 0.00000\ 3467\ T^2$ (0.003052) (0.00000066)

(2) $AVC = 226.06 - 0.01044\ T + 0.000003523\ T^2$ (0.002337) (0.0000005114)

$$R^2 = 0.8406$$

The minimum cost point appeared is at 1455 tonnes and 1482 tonnes of installed capacity per day respectively, for the ATC and AVC cost curves. From this analyses, it appears that both ATC and AVC decreased earlier and then increased with the installed capacity.

(1977)

Pawar while studying economics of scale in co-operative sugar units in Maharastra concluded that per unit profitability of sugar production was low in large sized factories as compared to smaller onces. Large sized factories, however were at advantage in economising manufacturing expenses and overhead costs provided the cane price and excise duty remain the same for all factories. Further he stated that the large sized factories could derive economics of scale by utilising their capacities to a higher level. These factories were in loss due to insufficient supply of cane. According to him the Problem of insufficient supply of cane could be corrected by increasing area under cane as well as by increasing productivity of cane. Limitations in irrigation however brings about restraint on expansion of area under cane cultivation.

The only alternative left is, therefore, to increase sugarcane yield per unit of area.

(1977, pp.5-8)

Bajapai reported that Indian sugar industries have been severely hit by the problems like low yield, uneconomic size, and low efficiencies and high losses in production. Small size sugar factories having very low installed capacity, no planning or any scope for their future expansion or modernisation, were clustered in the areas of sugarcane availability.

(1978, pp.89-91)

A report compiled by Eid-parry Ltd. Tamil Nadu expressed that there is a loss of Rs. 4.11 crores due to lower working efficiency and less crushing duration. The factory could not be run efficiently due to mechanical defects and labour problems. The start of the sugar factories was delayed due to several adverse factors affecting the number of working days.

(1979, p. 115)

The study conducted by Baviskar relating to the Kisan Co-operative Sugar Factory in Maharastra is worth mentioning, wherein an attempt is made to analyze the interrelationship between the co-operative movement and politics. He conducted that there is a dynamic relationship between development and politics. The existing political structure and given distribution of power within them influence the course of development which in turn creates new opportunities and resources bringing about change in the very structure and style of politics in the area concerned. However the study refers to socio-political aspects of sugar industry in rural Maharastra.

(1980)

Mahalingam has indicated that co-operative sugar factories are one of the important co-operative processing industries in India. It brings in a lot of benefits to the farmers who supply the cane to the factory on a remunerative price. He is of the opinion that co-operatives should make a positive contribution by supplying various inputs to formers and recovering the same from the cane price in easy installments, thereby both could be benefited.

(1980, pp.107-113)

Ramesh stated that the sugar industry in India is the second largest processing industry in the country next only to textile. There are 320 factories producing annually 65.70 lakh tonnes of sugar. The aggregate assets of the industry are more than Rs. 1300 crores. About 30 million cultivators are engaged in growing sugarcane and supplying the same to sugar factories. The sugar industry distributes about Rs. 800 crores annually towards sugarcane price. It contributes more than Rs. 300 crore annually to state and central exchequer. The factories are ideally suited for faster rural socio-economic development. The prosperity of the sugar industry is, therefore, closely linked with the development and prosperity of our vast population resident in the villages.

(1980, pp.47-48)

Shah and Shah studied that the cost of production of sugar factory depends primarily on raw materials, sugar recovery percentage and the duration of the crushing season. They suggested that, the cost of sugar production can be brought down by utilising the processing unit for a maximum period, by proper checking of the machinery by reducing cost of extra fuel, lubricants, spare parts, consumption of chemicals and sugar content in final molasses. It must be ensured that the steam balance and machinery are maintained with proper plans and proper watch of boiling house stations.

(1980)

Mane stated that the sugar co-operatives had blazed a trail in efficiency only to accelerate economic improvement of the formers. There is a raise in their level of income, saving investment, mostly in productive assets, possession of farm machinery, credit worthiness and employment opportunities.

(1981, pp.47-48)

Muralidaran compared the processing of sugarcane into sugar, *gur* and *khandasari* in Mandya district of Karnataka. The establishments of cost of three processing units were in the order of Rs. 4,02,83,292.03 per sugar unit (of 1250 tonnes crushing per day), Rs. 46,329.83 for *gur* unit and Rs.9, 16,722.38 per *Khandasari* unit. Share of fixed cost in the total processing cost was 64.24 per cent in sugar units, 30.78 per cent in *gur* and 17.29 per cent in *khandasari* units, were as the variable costs formed at respective percentages of 35.76, 69.22 and 82.71 for sugar, *gur* and *khandasari* units.

(1981)

Shinde stated that the escalation in the cost of plant, machinery and civil works has been a major source of sickness and a new co-operative sugar factory requires four to five years to break-even. It incurs an annual loss of Rs.25 lakhs to Rs.40 lakhs until than the losses accumulate rapidly due to high interest and the unit falls sick.

(1981, P.5)

Athre found that the rate of return on capital investment was comparatively high in large sized sugar factories, the rate of return on capital investment was observed to be less than the market rate of interest even for productive period of 20 years. In case of the medium sized sugar factories, however it exceeded the market rate of interest provided the productive period remains at 15 years and above. The low capacity utilisation on account of inadequate availability of quantity cane within area of operation resulted in heavy financial loss as the cost of production is higher.

(1982)

Anonymous showed that India was perhaps the only country in the world with very big co-operative base in sugar industries. Out of the total installed capacity of 67.87 lakh tonnes of sugar production, the share of the co-operative sector was 36.68 lakh tonnes.

(1984, p. 167)

Chougule stated that to get good recovery, there should not be less than 40 per cent of total cane area in a factory zone under early maturing varieties, 30 per cent and 30percent. under mid and mediate maturing varieties. It has been reported in Anakapalle Research Station that climatic changes do not affect the juice quantity of early maturing varieties to a significant extent unlike in the case of other varieties. The extent of loss in quantity in early maturing varieties will however be much lower than others. Drought resistant and early maturing varieties may be preferred for drought prone areas. There is need for studying carefully the nature of sugarcane development needed in each factory zone by technical men to formulate a scheme of development. This has to be implemented by the management in a meticulous manner to derive maximum profit.

(1986, p. 45)

Jain revealed that a large number of joint stock sector sugar mills are concentrated in Uttar Pradesh. An average size sugar plant receives cane supplies from as many as 40,000 growers, who have small holdings and constitute a significant segment of our rural population as against this, the number of case growers is much smaller in Maharashtra and Gujarat where the sugar industry is mostly in co-operative sector. It has to be appreciated that economically backward areas with large concentration of sugarcane growers having small land holdings need to be given an essential priority. Whereas in Maharashtra and Gujarat the sugar industry is mostly in the co-operative sector given priority in the matter of short term loans for co-operatives is immediate essentiality for the sugarcane production.

(1987, p. 2)

Naidu and Shrinivasan pointed out that the sugarcane plays an important role in Indian economy. Presently sugarcane is the seasonal and the largest excise revenue-earning commodity. It is estimated that about 500 crores of rupees of excise duty is earned from about 80 lakh tonnes of white sugar produced in a year.

(1987, p.1)

Venkatesan has suggested that the decision of Government of India to have 2500 tonnes crushing per day as the minimum viable capacity for licensing new sugar mills has increased the initial cost of establishment of new sugar mills by almost doubling the total investment required.

(1987, p.7)

Narayanaswamy and Ramchandran analysed the growth and development of Amaravati Sugar Mill, Tamil Nadu with respect to 13 identical indicators such as (1) Area under sugarcane production, (2) Membership, (3) Recovery, (4) Equity capital, (5) Debt capital, (6) Net working capital, (7) Cane price, (8) Cost of production of sugar, (9) Machinery utilisation, (10) Sale price, (11) income, (12) Expenditure and (13) Profit, compound growth rate was calculated for each indicator to study the growth.

(1988, pp.243-245)

The study of Hinge *et. al.* showed that the installed capacity was over-utilised in the healthy class while in the remaining classes; it was underutilisation due to inadequate cane supply which in turn influenced per unit cost of production. The gap between the highest and the lowest per quintal cost of manufacturing sugar was Rs. 118.3. The average per annum value of sugar was the highest in the healthy classes followed by medium and sick sugar factories. The sugar factories belonging to all the classes incurred loss. However, the loss was the highest in the sick sugar factories. The net loss of 100 tones of installed capacity was observed to influence the magnitude of return from sugar production. In spite of the low per unit cost of production of sugar, the overhead costs, were relatively very high in the case of sick sugar factories. These sugar factories sustained heavy losses.

The economics of scale entirely depend on the ability of sugar factories to fully utilize the installed capacity.

(1989)

Kasar and Tilekar indicated that the sugar industry has significant impact on the employment of seasonal migrants in Maharastra. The share of sugar factory employment was to the extent of 65.36, 45.51 and 75 per cent in the total employment of an average male, female and bullock pair of the household, respectively. As regards the income, it is noted that the sugar industry, on an average contributes 57 per cent of the gross income of the household. The seasonal employment provided by the sugar industry enabled the migrants to increase their income to enjoy a slightly better position as compared to the non-migrants under study. Hence, the policy has been endorsed for the installation of agro-processing industries based on local raw material opportunities for the economic development of weaker sections.

(1989, p. 329)

The study of Herekar on the performance of ten selected sugar factories in Karnataka and Maharastra states revealed that the age and capacity utilisation had some positive impact on the financial performance, whereas crushing capacity and sucrose recovery did not have much association with their financial performance.

(1995, pp.33-48)

Tuteja *et al.* the committee opined that, the sugar industry found itself entangled in a complex web of problems of high stocks, low prices, poor profitability, mounting cane price arrears, financial crunch (or outright sickness), limited modernisation, expansion, diversification, and weak international competitive edge. Suitable steps by various stake holders (including State Governments) are necessary to ensure that sufficient sugarcane is developed and grown in the mill area for the economic viability of sugar factories. Scientific rationalisation of cane area will have a direct impact on the economy of the sugar industry. The Central Government may dispense with the release mechanism for free sale sugar with effect from 1st October 2005. Beyond the initial time limit, a maximum of 3 months may be permitted for lifting of levy sugar, where after, the levy sugar quota would automatically be converted into free sale sugar, without any recurring levy obligation on this portion of levy sugar. Indian Sugar Exim Corporation (ISEC), an organ of the sugar industry, may play a prominent role both in import and export of sugar.

The minimum radial distance between an existing sugar factory and new factory may be 25 km, in place of the present 15 km. Additional requirement of sugarcane for expansion of capacity beyond 5000 TCD should accrue through increased productivity and not by expansion of the area under sugarcane.A scheme similar to the one available for integrated textile units should be sanctioned for the sugar sector for reducing their debt service burden. In order to improve profitability of sugar industry through value addition to by-products like bagasse and molasses, co-operative sugar mills having potential for setting up of diversification projects may be encouraged. Rate of interest on existing loans from Sugar Development Fund (SDF) may be pegged at 2 percentage points below the prevailing bank rate. As sugar industry do not discharge any toxic or highly polluting wastes, it may be placed in the orange category.

The following policy intervention is necessary for exploitation of by-products:

(a) Oxygenation of gasoline may be made compulsory.

(b) The Central Government may strictly implement its long-term policy for blending of petrol with ethanol.

(c) Fiscal and other incentives provided (in regard to blending of petrol with ethanol) may be in place for a minimum period of five years, without any rollback.

(d) Purchase price of ethanol may be fixed for a period of three years (with suitable price escalation clauses) in place of the present practice of fixing the price annually.

(e) The Central Government may take steps to prevent frequent changes in PPAs for cogeneration projects.

(f) National Energy Policy may have provisions for (i) mandating the proportion of 'green' power to be purchased in the overall power purchase in a state; and (ii) preferential tariff for co-generated power. The Central Electricity Act may also be suitably amended to provide teeth to the policy to support 'green' power.

In order to ensure the Research and Development efforts get a proper direction and fillip in the country, the efforts of the existing institutions may be adequately strengthened for increasing productivity in the fields as also in the mills by introduction of new technologies and also by effective extension mechanism. Sugar Development Fund may also assist project linked extension activities of these institutions. SDF may suitably reimburse the cost of organising national seminars for propagating successful research undertaken by R&D institutions. The industry itself may, in future, take up research in sugarcane as well as in sugar process & engineering through specifically created autonomous organisations, who would be in a position to ensure quick dissemination of the research findings. For industry supported new R&D organisations, capital assistance for building and other equipment may be considered from SDF upto 50 per cent, subject to a limit of Rs. 2.0 crores.

(2004)

1.2 Ratio analyses

Michael explained the meaning of balance sheet and the method of analyses to understand the financial structure of the firm. He suggested two ratios to test liquidity of a business namely, the current ratio and the quick ratio. Quick ratio should always be greater than one, but this is not always essential to solvency. This is true of the current ratio also.

Further, he suggested that it was not necessary for both the ratios to move in the same direction. If liquid ratio falls, debts and cash are in decreasing trend, but stock will be in increasing trend. If current ratio is rising, it leads to the building up of stocks, It would be unusual to find the falling current ratio and a steady liquid ratio, unless stocks are deliberately being rundown.

(1980)

Anand employed the solvency ratios, stock to working capital ratio, profitability ratio, gross profit ratio and net profit ratio to evaluate the financial position and performance of State Consumers Co-operative Federation, Maharashtra.

(1980, pp. 263-267)

Kumar analysed the financial performance of a bank by applying certain financial ratios. He opined that the ratio analyses provides guidelines and clues for spotting trends towards better or poor performance of the institutions. He emphasised the success of the institution to attain a stable financial position. Further he suggested introducing ratio analyses on a monthly basis to effectively appraise the performance of the institution.

(1983, pp. 621-622)

Ananth evaluated the performance of the Grape Growers Marketing and Processing Co-operative Society in Bangalore from 1958-59 to 1980-81 by employing the ratios such as solvency ratios, liquidity ratios, profitability ratios and turnover ratios. The study revealed that the Management efficiency of the society has improved over the years. A large addition to fixed assets to owned funds ratio (9.0349) during consolidation phase was a desirable improvement. The study indicated that the society showed an impressive record of performance during the last four years of study. It was also revealed that the society achieved a ratio of 20 and 18 per cent during 1979-80 and 1980-81 with regard to net profits to fixed assets.

(1984)

Rayude indicated that ratio analyses gives the required information about financial operations and performance. He used current ratio to indicate the short-term solvency to meet current liabilities out of current assets. Raw materials, finished goods, work in progress, stores, packing materials, debtors, loans, advances, deposits, cash on hand and in banks and other current assets were includes in current assets. Current liabilities include creditors deposits, provisions and other current liabilities. Any mill did not achieve the standard sound norm of current ratio 2:1. He also had used acid test ratio, where none of the mill has a ratio of 1:1. Debt equity ratio was used to measure the relative proportion of borrowed funds and members' capital in financing the assets of the societies.

(1985, pp.54-70)

Nikam made an attempt to study the financial strength of four co-operative sugar factories situated in Aurangabad district. The important ratio *viz.* current ratio and acid test ratio were employed to locate financial strength of these units (short terms) and two ratios *viz.* debt equity ratio and net fixed assets to net worth ratio were used for assessing the long-term financial strength of the societies.

(1986, pp. 244-254)

Shankaramurthy evaluated the performance of MARKFED (Marketing Federation), Karnataka, in terms of both physical and financial indicators. Various financial ratios were used to evaluate the financial performance. He observed that there was a substantial increase both in physical indicators and financial indicators of the federation over the period of study (1960-61 to 1984-85).

(1986)

Thanulingam and Gurumoorhty analysed 30 handloom co-operative societies in Paramakudi town, Tamil Nadu and have employed three categories of ratios

(i) liquidity ratios-used to judge firms short term obligations in which he had employed current and quick ratios as a measure of liquidity.

(ii) Profitability ratios-profitability is a meaningful measure and can be used as an effective standard of performance, in which he had employed gross profit ratio, net profit ratio, and operating profit ratio,

(iii) Turnover ratios- to indicate the rate at which assets were turned over and in this case stock turnover ratio and debtors turnover ratios have been used. They concluded that heavy accumulation of inventory and large quantity of debts created a high current ratio and did not indicate the ability to meet short-term obligations. In the profitability side, gross profit margin was too little to meet further expenses.

(1987, pp. 132-142)

Panday and Bhat indicated the multiple dimensions of financial phenomena generally referred to as liquidity, profitability, activity and leverage. They tried to narrow down the set of financial ratios without much loss of relevant and significant information, which the decision maker required. Nearly 30 ratios were computed and eleventh factor representing unique financial dimensions were obtained. The factors considered by them were return on investment (profit before depreciation, interest and taxes, tangible assets), sale efficiency (profit after taxes and net sales) equity intensiveness (retained cash flow and tangible net worth), short term liquidity (quick assets / fund expenditure), current assets intensiveness (inventory / current assets), cash position (cash / current liabilities), activity (inventories / cost of production), earning appropriation (dividends / profit before tax), financial structure (total liabilities / total tangible assets), interest converge (profit before interest and taxes / interest) and long-term capitalisation (long-term debt / total long-term capital).

(1988)

Krishnaswami and Narayanaswamy indicated that co-operatives shall finance their assets with equity and a co-operative organisation with a high operating leverage shall resort to a higher equity base in its capital structure to minimize the risk. They observed that a higher propagation of share-contribution by the government would help the co-operative to build up a strong equity base and to ease the burden of debt.

(1990, pp.267-275)

Mattigatti while conducting a study on performance of milk producers co-operative societies and their impact on dairy farming in Dharwad district, Karnataka, made use of different financial ratios to study the financial position of the societies.

(1990)

Mohandas evaluated the success of Kerala State Co-operative Rubber Marketing Federation (KSCRMF). He aimed at

(i) Assessing the viability of selected primary rubber marketing societies and KSCRMF with reference to turnover in rubber trade and

(ii) To examine the degree of loyalty of the primary societies to the federation in respect of the procured rubber. He employed profit-volume ratio to assess the volume of turnover and viability.

(1991, pp 126-134)

Padmini and Lekha studied the financial performance of Shreė Naryana Co-operative Society by employing various financial ratios such as turnover ratio, liquidity ratio, sales ratio, etc.

(1992, pp.93-105)

Hegde evaluated the performance of the Karnataka Co-operative Oil seeds Growers' Federation. He employed solvency, liquidity, profitability, efficiency and turnover ratios for evaluating the financial performance.

(1993)

Hind carried out his study to determine whether agricultural co-operative companies in UK were more likely to under-perform when compared with non-co-operative companies in the same sector. The research indicated that on the basis of profitability, capital gearing and liquidity, there were no significant differences between the two groups. The study revealed that the co-operatives did not perform differently compared to non-co-operatives with respect to operational efficiency indicators.

(1994, pp.213-219)

Pai *et. al.* made an effort to study the relationship between diversified firms and their financial performance. A set of performance measures/ratios were employed to determine the level of financial performance subsequently Kruskal-Wallis test was used to rank firms to establish relative superiority of performance.

(1995, pp.977-988)

Narasaiah analysed cash management in Kovur Co-operative Sugar Factory, Nellore, Andhra Pradesh. He analysed cash and bank balances as percentage to the current assets, current ratio, liquid ratio and cash flow coverage ratio. The study indicated low cash and bank balances when compared to current assets, which resulted in very low liquid assets. Inventory occupied a major portion in current assets and the net cash flow coverage ratio of the factory was negative due to inconsistent and insufficient Maintenance of liquidity.

(1996, pp.268-214)

1.3 Compound Growth Rates

Verma studied the growth rate of Dairy Co-operatives in India for the period from 1963-64 to 1975-76 using linear regression equations. The variables considered were number of dairy co-operatives, membership, paid up share capital and value of sales during the year. In order to study the comparative growth of societies, the percentage annual growth rates were also estimated.

(1980, pp. 45-49)

Bhale Rao *et. al.* studied the growth of arecanut marketing societies in India for the period from 1967 to 1977. The analyses of year wise growth of areca nut marketing societies considered growth in number of societies, membership share capital, working capital, etc. The study indicated that, the growth of societies was not consistent and with their annual fluctuations.

(1972, pp.365-369)

Gracy studied the growth and performance of primary agricultural credit co-operative societies in different states of India for the period from 1959 to 1979. She used compound growth rate model in seven performance variables using exponential functions, for which 1970 was considered as the base year. The results indicated that the growth performance of PACS based on membership, share capital, working capital deposits position and the amount of loan advanced for society increased gradually whereas the growth rate was higher during the post green revolution period.

(1983)

Subbarao analysed the growth rate of certain variables like share capital, membership, total assets in his study on business performance of the Central Areca nut Marketing and Processing Co-operative Ltd., Mangalore, Karnataka from 1973-74 to 1980-81. The growth functions of both exponential and modified exponential form as given below were fitted to a certain performance variable.

i) $Y = ^{abt}$ – exponential function

ii) $Y = ab\ t\ y\ c^{t2}$ – modified exponential function

Where,

Y = indicator, a = constant, b and c = regression co-efficient

t = years (time).

He observed that annual rate of increase of different performance variables were not uniform. The growth of paid up share capital, membership, total assets, working capital and purchases increased steadily, while in case of sales, the growth rate was high in the initial years and declined in later years (from 1977-78).

(2000)

Shankarmurthy used compound growth rate analyses of exponential form while studying the performance of Karnataka Co-operative Marketing Federation. The important physical indicators considered were membership, branches, depots, employees, direct recruited and deputed officials. The important financial indicators were share capital, owned funds, fixed assets, total assets long-term investments, total working capital, total liabilities, inventories, total sales, sale of fertilizers, sale of other commodities and establishment expenses. In case of financial indicators, constant price was followed for the sub-periods and for the aggregate period to account for the inflationary trend in the economy in order to get real picture of the situation.

(2001)

Kumar in his study of performance of dairy co-operative and their impact on milk production, income and employment in Chittoor district, Andhra Pradesh, used the compound growth rate analyses for various physical and financial indicators or the selected milk producers' co-operative societies. To estimate the growth rates employed exponential function of the following type.

$Y = ab^t$

Where,

Y= indicator, a= constant, b= regression co-efficient, t= years (time)

(2002)

Gupta used compound growth rate mode of exponential form to study the performance of agricultural credit co-operative societies in Shimla district of Himachal Pradesh.

Main emphasis was given on the parameters such as, total loan advanced, short term and medium term loans advanced, cash and kind components of loan advanced, loan advanced to different categories of farmers, recovery position and overdue of loans advanced by the societies in the district.

(2003, pp. 241- 243)

Kumar used growth rate analyses to analyze the growth in physical and financial performance indicators of Horticultural Producers Co-operative Marketing Society Limited, Bangalore. The indicators considered were membership, share capital, owned funds, sales, inventories; fixed assets, current assets, total assets, current liabilities and total liabilities.

(2004)

Babureddy and Chengappa evaluated the progress and performance of Primary Co-operative Agricultural and Rural Development Banks in different districts of Karnataka. The compound growth rate model of exponential form was estimated for different physical and financial indicators.

(2004, pp.129-138)

1.4 Performance of Agri-related institutions and agro-based Industries

Tiwari and George conducted an opinion survey on marketing of agricultural produce through co-operative marketing societies in Punjab and Haryana states regarding the factors influencing farmers choice of special agencies in disposing off their produce. It was found that (i) private traders had an edge over co-operative marketing societies (ii) cultivators did not sell all their products through co-operatives, even though they were members and (iii) Private traders were successful compared to the co-operative because of the absence of complicated and legal procedures in their transactions and the additional facilities provided by them to their customers such as liberal loans, free storage, transport, supply of consumer goods and farm inputs on credit, personal relations and hospitality.

(1970, pp. 206-214)

Dhankar evaluated the performance of a co-operative bank in Rajastan in terms of their owned funds, deposits, savings, borrowing, grants, loans, over dues, investment, and other service facilities to its customers. He suggested that co-operative should adopt a progressive attitude towards its activities and offer a wide range of facilities to its members for its success.

(1971, pp. 61-80)

Bhalerao and Vishwanath in their study on co-operative marketing of fruits and vegetables, highlighted the weaknesses of the societies and stressed that they should be urgently removed if they had to create a significant impact on producers in order to enroll a majority of fruit and vegetable growers as members. An intensive drive would have to be launched. Further, they recommended for state partnership by way of contribution to share capital as well as loans and subsidiaries for management, storage, and transportation and processing equipment.

(1972, pp. 365-369)

Mariyanthan evaluated the progress and performance of the Primary Agricultural Credit Societies in India, considering the credit (which included purpose and type of loans) and non-credit operations. Facilities for marketing and processing of agricultural produce of the members were also considered for the analyses. Further, he concluded that the larger membership per society added to the strength of its business activities and opined that the membership and capital position were the two necessary ingredients that would lead to the success of any co-operative society. Later, he concluded that the societies were not providing adequate facilities for marketing and processing of agricultural produce of the members, which constituted a great lapse on their part.

(1973, pp. 465-471)

Gupta and George made a study on modernisation of rice processing industry in Punjab. In their study they found that the fixed cost per mill increased with an increase in the size of the mill. The fixed cost was higher for modern mill than for traditional mills. The average variable cost for all the mills together was Rs.2.33 per quintal of paddy milled. And also they stated that the average cost curve was 'U' shaped curve where unit cost declined as the volume increased up to a certain volume level and then it increased. Further, the fixed cost per tonne declined as the capacity increased from half tonne to one tonne. They stated that an efficient mill is one, which gives maximum returns per quintal of paddy milled. Their contentions was that since return is measured in terms of value of output minus costs, the mill with the minimum cost per quintal may not necessarily be the optimum size.

(1974)

Das made a detailed study of economics of sunflower cultivation and processing in Tamil Nadu. Their study revealed that processors get 12.76 per cent of consumer's rupee spent on oil and cake and farmers got 82.86 per cent. They estimated average processing cost at Rs. 165 per tonne of seed with each tonne of seed costing Rs. 239. But the average oil yield was 27.5 per cent and thus the seed cost per tonne of oil worked out to Rs. 3, 636. The average price realised was Rs.8, 030 per tonne for oil and Rs.833 per tonne for cake and thus the processors margin per tonne was Rs. 718. They stated that the processors often faced the problems in the procurement of sunflower seeds, which had an effect on their profit margins.

(1975)

Anonymous in the project for Integrated Development of Horticulture in Karnataka stressed the importance of processing industry for utilising fruits and vegetables during the glut season for processing, thereby regulating the prices in the market and also for making available the seasonal fruits and vegetables during the off season. They observed that out of the 0.025 million tones of existing installed capacity per year, utilisation was only 0.012 million tonnes. The project has also suggested measures for strengthening and streamlining of existing units in addition to the proposal for starting new units.

(1976)

Rao evaluated the co-operative movement in Andhra Pradesh in terms of membership, capital resources, lending operations, facilities and services provided by the society and distribution of consumer goods and farm requisites. He also considered profitability, over dues, audit classification, loans issued and the outstanding and the management and its cost for assessing the economic viability of the co-operative societies.

(1977, pp. 239-240)

Karche mentioned that the agricultural co-operative processing societies are the organisations of farmers, formed for processing agricultural produce for better returns to the farmers. The success or failure of the co-operatives depend on the availability of raw materials. He discusses the advantage of co-operative processing like savings in transportation cost, etc. The member's failure to supply raw materials leads to under utilisation of capacity and thereby increasing losses, wiping away of share capital and finally if steps are not taken, leading to the liquidation of the organisation. He suggests that for the survival and growth of the organisation, it must have the effective policy of raw material procurement from the members.

(1978, pp.17-26)

Muralidharan made a comparative study of processing sugarcane into sugar, *gur* and *khandasari*. He found that nearly 63.23 per cent of processing costs formed fixed items in sugar units and this was mainly because of inclusion of salaries and wages of permanent staff. In case of *khandasari* unit, fixed costs constituted 17.29 per cent and variable cost 82.71 per cent. While in case of *gur* unit fixed costs accounted for 30.78 per cent and variable costs 69.22 per cent. In case of *gur* unit labour cost was directly linked to the quantity of *gur* produced indicating that an increase in the quantity of *gur* would reduce the per unit cost. He found processing cane into *gur* was more profitable when compared to sugar and *khandasari*.

(1981)

Gracy in her study on the progress and performance of Primary Agricultural Co-operatives Societies in selected States of India has considered the following as indicators of performance such as (i) membership and coverage (ii) financial resources (iii) credit activities (iv) cost of management and profit and loss and (v) distribution of credit to the weaker sections particularly to schedule castes and tribes.

(1983)

Acharaya found that 78 per cent of processing cost of milk was accounted for raw milk and rest 22 per cent constituted skim milk powder and other miscellaneous expenses. He reported that 55.4 per cent was accounted for fixed cost in the total cost of processing per litre of standardized milk while variable cost was 44.6 per cent.

(1984)

Gajja concluded in his article that cost of collection has lion's share in total cost of milk processing. The cost of raw milk showed an increasing trend but the cost of transportation showed a decreasing trend. Further, the results revealed that procurement of mill was increasing but at a decreasing rate much below the capacity of the milk plant. The procurement of milk during 1977-78 was 226.5 and 175.81 lakh litres for Jodhpur and Bikaner milk plants of Rajasthan respectively as against installed capacity of 365 lakh litres. The study showed that there were ample scopes for milk procurement for the two milk plants studies.

(1985, pp. 82-93)

Shankaramurthy studies the performance of Karnataka State Co-operative Marketing Federation Limited. He used different ratios to study different aspects of financial position of the federation such as solvency, liquidity, turnover, profitability, efficiency and strength. He said that the ratio analyses would provide better idea of the financial position of the federation.

(1986)

Mohammad and Raghuram observed that cashew processing as an important agro-based industry with a potentiality of employing about 2.8 lakh persons, mainly women. Labour cost accounted for a major share nearly 56.6 per cent in the total cost of processing.

(1990, pp. 187-189)

Jawanram made an attempt to analyze the organisation and working of Jaipur District Milk Producers Co-operatives Union Limited, Jaipur. The study was conducted through personal interview with management and other employees of the union. It was found that the organisational structure and functions performed such as (i) milk collection (ii) supply of technical inputs (iii) farmers induction programmes and (iv) supervision were analysed. Some drawbacks were found out and appropriate suggestions were made.

(1993, pp. 273-282)

Srivastava indicated that with subsequent secondary and tertiary processing of various raw materials the value added as well as the price of finished products would be increased. He observed that agro-processing units accounted for 39 per cent of all factories, 12 per cent of fixed capital, and 13 per cent of working capital and 15-per cent of total capital employed in the industry in the organised sector. This 15 per cent of total capital employed in the industry in the organised sector. This 15 per cent of capital investment generated 3 per cent of the total employment, 26 per cent of the output and 21 per cent of the net value added. He noticed capital productivity in agro-industries was almost the double (0.7) than that of non-agro-based industries (0.35). While labour productivity was less than half when compared to non-agro-based industries.

(1995, pp. 12-15)

Mattigatti studied the performance of Milk Producers Co-operatives Societies and their impact on dairy farming in Dharwad district. The researcher selected number of physical and financial indicators to evaluate the performance. The secondary data required was collected from the various annual reports of Milk Producer's Co-operatives Societies for the period 1986-88. He opined that both the physical and financial indicators of the societies showed significant growth in their values. The above average societies had already progressed with higher values for the indicators compared to below average societies, while below average societies had shown a faster rate of growth. Hence, he concluded that over the period of time all these societies would contribute to the overall development of the area.

(2000)

CHAPTER 2

OVERVIEW OF SUGAR INDUSTRY IN INDIA

Indian sugar industry, the second largest agro-based processing industry after the cotton textiles industry in the country, has a lion's share in accelerating industrialization process and bringing socio-economic changes in under developed rural areas. Sugar industry covers around 7.5 per cent of total rural population and provides employment to 5 lakh rural people. About 4.5 crores farmers are engaged in sugarcane cultivation in India. Sugar mills in co-operative, private, and public sectors have been instrumental in initiating a number of entrepreneurial activities in rural India. Present work is an attempt to review the progress of sugar industry in India, to understand its problems and challenges in the context of ongoing liberalization process. Indian sugar industry can be a global leader provided it comes out of the vicious cycle of shortage and surplus of sugarcane, lower sugarcane yield, lower sugar recovery, ever increasing production costs and mounting losses. It needs quality management at all levels of activity to enhance productivity and production. Attention is required on cost minimization and undertaking by-product processing activities.

Five-year Plans and Sugar Industry

The sugar industry was granted protection till 1950. Since independence there has been an overall increasing trend in sugar production in India. Production of sugar has increased by leaps and bounds in the planning period. To meet the increasing sugar requirement during different plan periods targets of sugar production were fixed as depicted in the Table 2.1.

Table 2.1

Progress of sugar industry in India under Five-year plans

Plan period	*Production Targets (Lakh tonnes)*	*Actual Production (Lakh tonnes)*	*Number of Sugar Mills*
First Plan (1951-56)	18	19.34	138
Second Plan (1956-61)	25	30.29	175
Third Plan (1961-66)	35	35.32	200
Fourth Plan (1969-74)	47	39.50	229
Fifth Plan (1974-78)	54	58.42	298
Sixth Plan (1980-85)	76	61.78	356
Seventh Plan (1980-85)	102	109.90	414
Eight Plan (1992-97)	143	139.56	412
Ninth Plan (1997-2002)	148	185	434
Tenth Plan (2003-08)	156	170	461

Source: Yojana Nov. 1999, ISMA Report 2004, 2005 and 2006.

Before the commencement of First Plan there were 138 sugar factories with an installed annual sugar production of 19.34 lakh tonnes. During the plan period, to achieve targets of sugar production, licenses were issued for setting up of new factories and for many of the existing units to expand the size of the units. The number of the sugar factories increased to 138 in the first plan, 175 in the Second Plan. The production increased to 30.29 lakh tonnes in the Second Plan. During Second Plan the target of production was 22.5 lakh tonnes, which was increased to 25 lakh tonnes, but the actual production exceeded up to 30.29 lakh tonnes, which was slightly more than demands. This resulted in decontrol upto some extent. In the Third Plan the target of production was

35 lakh tonnes. Due to shortfall in production of the cane in first three years of the Third Plan the target could not be fulfilled but at the end of the Plan the target of production up to Third Plan was more than target but due to seasonal variations the target could not be achieved in fourth plan. Again in Fifth plan the production was more (58.42 lakh tonnes) than target (54 lakh tones). In the Sixth Plan the target was 76 lakh tonnes but the production was only 61.76 lakh tonnes. Again in Seventh Plan it was more than target. In Eight plan the target further could not be achieved. Although the production of sugar decreased in 1992-93 and 1993-94 but it increased to 146 lakh tonnes in 1994-95 and India became largest sugar producing country in the world. In 2002-2003 the production of sugar in India was 28 lakh tonnes, which decreased to 170 lakh tonnes in 2003-04. In 1950-51 there were 138 sugar mills in India but upto 31st March 2007 this number increased to 461. At present there are 553 registered sugar factories having capital investment of Rs. 50,000 crores and annual production capacity of 180 lakh metric tonnes and presently sugar industry is the second largest agro-based industry of India.

Production of Sugarcane and Sugar

Both area and production of sugarcane fluctuate considerably from year to year. Table 2.2 shows progresses of sugarcane area, production and yield from 1994-95 to 2005-06. The fluctuations are due to variations in climatic conditions, the vulnerability of areas cultivated under rain fed conditions, fluctuations in prices of gur and khandasari, and changes in returns from competing crops. Despite this instability, both area and production of sugarcane have increased considerably over the past three decades. The chief raw material for sugar production in India is sugarcane. The data shows that area under sugarcane and production had shown a mixed growth during 1994-95 to 2005-06; of twelve years, seven years show positive growth rate and Five years show the negative. As far as sugarcane production is concerned, it had negative growth during for six years due to lower rainfall. There had been considerable variations in area under sugarcane, sugarcane production, and yield of sugarcane during 1994-95 to 2005-06.

Table 2.2

Progress in Sugarcane area, production and yield from 1994-95 to 2005-06

Year	*Area ('000' ha)*	*Production ('000'tonnes)*	*Yield/ha*
1994-95	3867	275540	71.30
1995-96	4147 (7.24)	281100 (2.01)	67.80 (-4.91)
1996-97	4174 (0.65)	277560 (-1.25)	66.50 (-1.92)
1997-98	3930 (-5.84)	279541 (0.71)	71.10 (6.92)
1998-99	4055 (3.18)	288722 (3.28)	71.20 (0.14)
1999-00	4223 (4.14)	299324 (3.67)	70.90 (-0.42)
2000-01	4316 (2.20)	295956 (-1.13)	68.60 (-3.24)
2001-02	4430 (2.64)	297208 (0.42)	67.40 (-1.75)
2002-03	4361 (-1.55)	281575 (-5.26)	64.60 (-4.15)
2003-04	3995 (-8.39)	236176 (-16.12)	59.10 (-8.51)
2004-05	4541 (-1.55)	299145 (-5.26)	64.8 (-4.15)
2005-06	4201 (-5.63)	281172 (-6.78)	66.9 (-5.78)

Source : Report of the committee on revitalisation of sugar industry, Government of India, Dec. 2004, Co-operative sugar, Vol. 38, August 2007.

Note : Figures shown in brackets indicate percentage change over previous year.

Fluctuations in sugarcane prices also led to under or over production of sugarcane during these years. Sugarcane yield had negative growth during eight years out of 12 years. As a result, sugar industry had shortage of sugarcane. Sugarcane yield per hectare in India was 59.10 tonnes during 2003-04. It is too low in comparison with sugarcane yield in Kenya (100 MTs), Hawai (150 MTs), and Australia (75 MTs). In view of this situation, India needs to increase sugarcane yield at par with global competitors to assure viability of sugarcane cultivation and sugar industry. Government should provide irrigation facilities, adequate, and regular power supply to agriculture sector. In India, 65-70 per cent of available sugarcane is used for manufacturing sugar, and 30- 35 per cent for making *gur* and *khandsari*. During 1994-95, there were 408 operating sugar factories in India. The number of sugar factories rose to 455 over period of 12 years, registering an increase by 12.99 per cent, which is shown in Table 2.3. Over the period of 12 years, sugar production saw substantial variations. During 1994-95, sugar produced by 408 factories was 14.64 million tonnes and during 1998-99 the production of sugar was the highest (20.92 per cent).

Table 2.3

Number of operating sugar factories, production, and recovery of sugar in India

Year	*No. of operatingsugar factories*	*Sugar production ('000 tonnes)*	*Recovery (%)*
1994-95	408	14643	9.92
1995-96	416 (1.96)	16543 (12.97)	9.42 (-5.04)
1996-97	412 (-9.06)	12905 (-21.99)	9.90 (5.09)
1997-98	400 (-2.91)	12852 (-0.41)	9.95 (0.50)
1998-99	427 (6.75)	15541 (20.92)	9.86 (-0.90)
1999-00	423 (-0.94)	18200 (17.11)	10.20 (3.45)
2000-01	436 (3.07)	18511 (1.71)	10.48 (2.74)
2001-02	434 (-0.46)	18529 (0.09)	10.27 (-2.00)
2002-03	453 (4.37)	20100 (8.48)	10.36 (0.88)
2003-04	422 (1.77)	17000 (-15.42)	10.22 (0.87)
2004-05	400 (1.09)	12690 (-11.45)	10.17 (-2.34)
2005-06	455 (3.73)	19267 (6.37)	10.21 (-2,34)

Source: Report of Indian sugar mills association, 2004,2005 and 2006.

Note: Figures shown in brackets indicate percentage change over previous year.

The data shows that there was negative growth of sugar production during 1996-97, 1997-98, 2003-04 and 2004-05. As far as sugar recovery is concerned, it remained below 10 per cent during 1994-95 to 1998-99, and slightly above 10 per cent during 1999-00 to 2005-06. Use of low yield varieties, unscientific harvesting of sugarcane and delays in crushing are some of reasons responsible for low sugar recovery. Global competitors, such as Australia have sugar recovery between 14.25 per cent, 16 per cent in Kenya, Columbia and Hawai have above 12 per cent. In the context of competitive global scenario, Indian sugar industry urgently needs to undertake research and development activities to increase sugar recovery.

Sugar Industry as Raw Material Supplier

Sugar industry is a major supplier of molasses for alcohol and ethanol producing units. Bagasse is used for power generation as well as paper production. Ethanol can be used as fuel. The data given in Table 2.4 shows production of molasses and ethanol during 1994-05 to 2005-6. Sugar industry produced 64.97 thousand tones of molasses and 12147 thousand litres of ethanol. There had been ups and downs in production of molasses and ethanol over period of 12years of study. Molasses production had decreased by 28. 35 per cent in 1996-97, 5.54 per cent in 1997-98 and 2.49 per cent in 2000-01 over preceding years.

Ethanol production and its consumption had a moderate increase. Ethanol consumption increased from 9000 thousand litres in 1994-95 to 15632 thousand litres in 2005-06. The ethanol consumption increased by 40.70 per cent during decade. At present, oil refineries are allowed to add 5 per cent ethanol in fuel. It may be increased to 10 per cent in future, which is essential to enhance economical viability of ethanol producing units.

Table 2.4

Production of molasses, ethanol and consumption from 1994-95 to 2005-06

Year	*Molasses ('000' tonnes)*	*Ethanol ('000' tonnes)*	*Ethanol consumption ('000' tonnes)*
1994-95	64.97	12147	9000
1995-96	82.85 (27.52)	15491 (27.53)	9450 (5.00)
1996-97	59.36 (-28.35)	10669 (-31.13)	9922 (4.99)
1997-98	56.07 (-5.54)	12632 (18.40)	10418 (5.00)
1998-99	69.76 (24.41)	13202 (4.54)	11486 (5.00)
1999-00	80.22 (15.00)	13801 (5.45)	10939 (5.00)
2000-01	78.22 (-2.49)	14427 (5.54)	12060 (4.99)
2001-02	80.66 (3.12)	15085 (4.56)	12663 (5.00)
2002-03	88.79 (4.57)	16321 (5.49)	13980 (5.67)
2003-04	59.05 (4.46)	16980 (5.94)	14235 (5.39)
2004-05	55.13 (4.42)	17231 (6.03)	14578 (5.89)
2005-06	85.49 (4.53)	17990 (6.85)	15632 (6.21)

Source : Report of Indian sugar mills association, 2004,2005 and 2006.

Note: Figures shown in brackets indicate percentage change over previous year.

Utilisation of Sugarcane for Different Purposes

The utilisation of sugarcane for production of white sugar, *gur*, *khandasari* seed, feed and chewing purposes for the period 1994-95 to 2005-06 is shown in the Table 2.5. The table shows that there is a considerable variation of sugarcane usage between the years. The contribution of white sugar production plays very important role in total utilisation of sugarcane in all the years. The considerable portion of the sugarcane is utilised for sweeteners like *gur* and *khandasari*. The percentage of sugarcane used for production of *gur* and *Khandasari* varies between 20.7 per cent during 2002-03 to 41.9 per cent during

1997-98. Small percentage of sugarcane is also used for seed, feed and chewing purposes.

Table 2.5

Utilisation of sugarcane for different purposes from 1994-95 to 2005-06

Year	*Production of sugarcane ('000'tonnes)*	*Sugarcane used for ('000' tonnes*		
		Production of white sugar	*Gur and Khandasari*	*Seed, feed and chewing*
1994-95	275540	147643 (53.6)	100248 (36.4)	27649 (10.0)
1995-96	281100	174726 (62.2)	73373 (26.1)	33001 (11.7)
1996-97	277560	130379 (47.0)	114172 (41.1)	33009 (11.9)
1997-98	279541	129135 (46.2)	117221 (41.9)	33186 (11.9)
1998-99	288722	157561 (54.6)	97692 (33.8)	33469 (11.6)
1999-00	299324	178494 (59.6)	86450 (28.9)	34380 (11.5)
2000-01	295956	176660 (59.7)	85366 (28.8)	33930 (11.5)
2001-02	297208	180346 (60.7)	82138 (27.6)	34724 (11.7)
2002-03	287383	194365 (67.6)	56494 (20.7)	33524 (11.7)
2003-04	233862	132511 (56.66)	61430 (26.27)	39921 (17.07)
2004-05	237088	124772 (52.63)	71843 (30.30)	40473 (17.07)
2005-06	281172	188672 (67.10)	68380 (24.32)	24120 (8.58)

Source : Report of Indian sugar mills association, 2004,2005 and 2006, Co-operative Sugar Vol.38, march 2007

Note: Figures shown in brackets indicate percentage change over previous year.

Export and Import of Sugar

Indian sugar industry contributes 15 per cent of global sugar production. While its share in global sugar consumption is around 13.4 per cent. Sugar exports and imports of India are shown in Table 2.6 during 1994-95 to 2005-06. However, exports of sugar declined substantially during 1998-99 and 1999-2000 by 1.93 and 1.09 per cent respectively over their previous and succeeding years due to decreased sugar production and uncompetitive sugar prices. Whenever there is lower production of sugar in the country, sugar is imported to meet domestic demand. Sugar imports during 1996-97, 2001-02 to 2003-04 had mixed growth trend.

Table 2.6

Export and Import of Sugar from 1994-95 to 2005-06

Year	*Sugar Exports (Tonnes)*	*Sugar Imports (Tonnes)*
1994-95	39350 (2.18)	1765440 (97.82)
1995-96	434320 (74.25)	150630 (25.75)
1996-97	667274 (99.68)	2131 (0.32)
1997-98	173282 (33.31)	346905 (66.69)
1998-99	17735 (1.93)	900471 (98.07)
1999-00	12990 (1.09)	1181183 (98.91)
2000-01	338691 (91.76)	30404 (8.24)
2001-02	1456448 (98.21)	26578 (1.79)
2002-03	1662370 (97.57)	41430 (2.43)
2003-04	1200600 (94.16)	74400 (5.84)
2004-05	108587 (10.43)	933000 (89.57)
2005-06	316850 (36.18)	559000 (63.82)

Source: 1. DGCI&S, Kolkata, 2006 and 2007
2. Report of Indian sugar mills association, 2004, 2005 and 2006.
3. F.O. Lichts 2004, 2005, 2006 and ISO Quarterly Market Outlook, November 2006

International Sugar Scenario

Sugar is produced in 110 countries. The leading sugarcane producing countries are Brazil, India, Australia, Thailand, China and Cuba. Sugar is extracted from two different raw materials, sugarcane and beet. Both produce identical refined sugar. Sugarcane is grown in semi-tropical regions, and accounts for around two-thirds of world production. Beet is grown in temperate climates, and accounts for the balance one third of world production. The Russian Federation, Ukraine and Europe account for around 80 per cent of total beet sugar production. In addition to weather conditions, diseases, insects, and quality of soil, production of sugarcane and beet are affected by international trade agreements and domestic price support programmes. The world production of sugar in 2005-06 was 142.33 million tones and sugar consumption was 139.76 million tones.

Critical Assessment

Sugar Industry occupies an important place among organised industries in India. Its main raw material is sugarcane. The special thing for all kinds of the raw material is that it should contain the highest percentage of the content for which it is used as raw material. But the quality of sugarcane of our country is not so good and researchers are trying to update it but due to lack of interest and proper attention from the government side, the efforts are not successful till today. Per hectare production of the sugarcane is also not improving which is a matter of great concern. It is probably due to lack of land fertility and lack of irrigation facilities. Other countries of the world such as Cuba, Fizzy and West Indies which are very small in comparison to India, are producing much more sugarcane per hectare than India. The share of India in the total production of sugarcane in the world is 37 per cent. But the production of sugarcane is only 15 tones per acre whereas in Java it is 56 tones and in Hawaii it is 52 tones i.e. almost four times than India. The production cost of sugar is also high due to inferior quality of Indian sugarcane. Since sugar mills are running at loss so they are unable to pay the cane growers on time. So the quality improvement in cane grower is the need of the time. The sugar policy of the government has been seriously lacking a long-term perspective. Controls, decontrols, partial controls, etc. have been used in past in an adhoc manner. It is necessary to assure supply of sugar to poorer sections at reasonable rate. But government policy on cane prices, control of price of sugar, dual pricing, etc. have been designed and implemented for the benefit of sugar mill owners and distributors and rarely for benefits of cane growers or for benefit of consumers of sugar. Much of the illness and problems of sugar industry are the result of the government's policy. In the sugar industry several by-products especially bagasse and molasses are found. Once bagasse was used as fuel and molasses were left without being used causing pollution and health hazards even today these by-products are neglected in some sugar factories due to lack of facilities. Small cottage industries may be established for disposing these by-products in a positive way for preparing paper, cardboards, alcohol, fertilizers, cattle field, etc. Apart from it manufacturing process of sugar also needs certain modification so that yield may be improved.

Challenges for Sugar Industry in India

Indian sugar industry has been facing crunch of raw material, and resource as well as infrastructure problems. Globalization has brought a number of opportunities but at the same time posed certain challenges before sugar industry. Most of sugar units in India utilize production capacity below 50 per cent. Low capacity utilization and inadequacy of raw material led to closer of 100 sugar factories in India. Mounting losses and decreasing net worth of sugar factories have been responsible for sickness of sugar industry. Sickness in sugar industry has reached to an alarming proportion. Indian sugar industry has been cash striven for decades. Low cash inflow due to piling stocks leads to serious financial crisis and finally to closing sugar factories. Sugar prices have been a political issue rather than economical issue. Many a times it worsens economy of sugar factories.

The main concern of sugar industry in India is fluctuations in sugarcane production due to inadequate irrigation facilities, lower sugarcane yield, and frequent droughts in tropical and sub-tropical areas where sugarcane is grown on large scale. In addition, sugarcane yield has been lower. Sugar recovery is also lower in comparison with other sugar manufacturing countries. This leads to escalation of production costs and weakness in giving competitive edge to the industry. Most of sugar mills in India are having daily sugarcane crushing capacity of 1250 tonnes. These mills cannot have economies of scale so they have to incur high production costs. Indian sugar industry is characterized by high production costs. Therefore, daily crushing capacity should be extended to 2500 tonnes. Obviously, industry has a great challenge of existence in global market. In recent years, sugarcane production in India has decelerated to a great extent due to water and power shortage. Special attention is needed to be given on water resource management. All the area under sugar cultivation should be brought under drip irrigation to conserve water as well as fertilizers. Adequate and regular power supply to sugarcane growers and sugar factories would increase production and productivity. To enhance share of Indian sugar industry in global trade, quality and quantity of sugar needs to be enhanced.

CHAPTER 3

SUGAR INDUSTRY IN KARNATAKA

Sugar industry is the major agro-based industry in Karnataka. Sugarcane cultivated for many decades as a major commercial crop in Karnataka. First sugar factory in Karnataka started in 1933-34 as public sector undertaking in Mandya district. The agro climatic condition in Karnataka is highly congenial for higher sugar cane yield and recovery. Sugar industry has given employment to the extent of 35 lakh people directly and indirectly in rural areas of Karnataka.

Karnataka sugar industry ranks third in terms of its contribution of sugar in the total sugar production in the country. The sugar industry in Karnataka is able to manufacture sugar in such huge quantities due to the fact that sugarcane is abundantly available in the state. In fact, Karnataka stands fourth in the country in the cultivation of sugarcane. In Karnataka there are two sugarcane-crushing seasons viz., Southern Karnataka – June/July to February/March, and North central and Coastal Karnataka- October/November to March/April.

Total area under sugarcane cultivation during 2002-03 was 3.85 lakh hectares and it was reduced to 2.60 lakh hectares during 2003-04 and again it was reduced to 1.78 lakh hectares during 2004-05. Total sugarcane produced during 2002-03 was 325 lakh tonnes and it was reduced 162 lakh tonnes during 2003-04 and again it has reduced to 142.76 lakh tones during

2004-05. The average sugarcane yield during 2002-03 was 84 tonnes per hectares, it has reduced to 65 tones per hectares during 2003-04 and it has increased to 80 tones per hectare during 2004-05. Total number of sugar factories working during 2002-03 were 43; they were increased to 46 during 2003-04, again reduced to 35 during 2004-05. The crushing capacity of the established sugar units during 2002-03 was 208 lakh tones; it was increased to 222 lakh tonnes during 2003-04 and during 2004-05 it has decreased to 208 lakh tonnes.

By the end of 2006 there were 46 sugar factories, which are distributed all over the state. Sugar industry in Karnataka is benefiting the formers, mill owners and government in many ways like, it has generated many facilities in the state such as communication, employment, and transport. It has also benefited the state by helping in the development of the rural areas of the state by mobilizing the various resources of the villages. The Sugar Industry in Karnataka contributes around Rs. 36 crore per year to the state exchequer through central excise duty. It also contributes more than Rs. 900 crores in the form of turnover tax and sales tax to the state exchequer. The state government has set up the Karnataka Sugar Institute (KSI), which has emerged as a centre for education and training for sugar technology. The Karnataka Sugar Institute also provides important support to the sugar industry in Karnataka by doing research and development in the various aspects of sugarcane processing and production.

Sugar industry in Karnataka has contributed a great deal to India's total level of sugar production and thus has helped the country to meet its demand for sugar. By 2006, in Karnataka, there were 41 sugar factories functioning of which 14 were functioning under co-operative sector, (forming 34 per cent of the total) followed by 25 factories under private sector (64 per cent of the total) and two factory under public sector (two per cent of the total). The recovery of sugar during 2003-04 was 10.59 per cent, it was reduced to 10.43 per cent during 2003-04 and it has decreased further to 10.11 per cent during 2004-05. In Karnataka nine sugar factories are having distilleries (four-co-operative, four-private and one public) and 5 industries are having distilleries with ethanol manufacturing units (production capacity of 130 KL/day). There are 10 sugar industries (private-8, co-operative-2) having cogeneration units with installed power generation capacity of 213 MW. The scenario of sugar industry in Karnataka by the end of 2007 is given below:

Sl. No	*Particulars*	
1.	Total number of sugar factories	52
2.	Total area under sugarcane (lakh ha)	3.19
3.	Average sugarcane yield (tonnes per ha)	89
5.	Total crushing capacity: Perday- Per season-	165750 tonnes 298 lakh tonnes
6.	Sugarcane crushed (lakh tonnes)	177.00
7.	Sugarcane production (lakh tonnes)	18.77
8.	Sugar recovery (%)	10.59
	By-Product Units	
9.	Molasses (kilolitres)	8.0 lakh
10.	Sugar factories having Distlleries	9
11.	Installed capacity of Ethanol production (kilolitres per day)	160
12.	No. of factories with Cogenration units	10
13.	Total power genteration	668MW

Need for the Study

The present work analyses the development and financial performance of the sugar industry in Karnataka. The purpose of this analyses is to identify influencing factors that are internal and external to this industry with a view to improving its performance and contributions to rural-led economic development.

Map showing the location of sugar factories in Karnataka

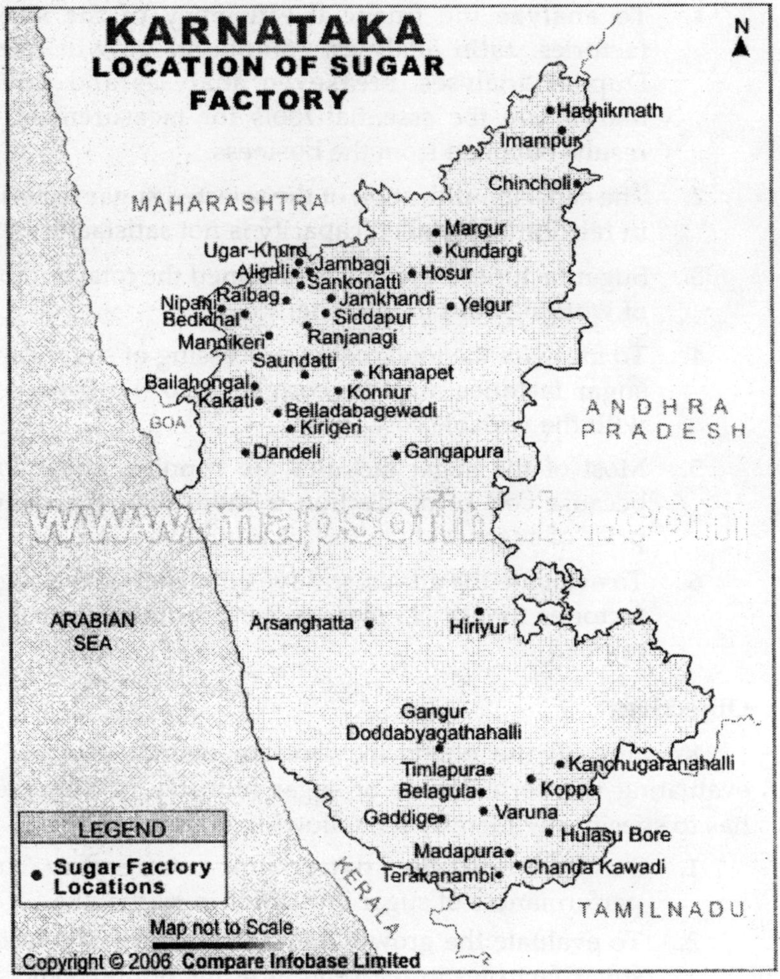

Source: www.mapsofindia.com

Hypothesis

The important hypothesis drawn here are:

1. To analyze the financial efficiency of the sugar factories, ratio analyses, compound growth rates, Dupont analyses, breakeven analyses and safety margin are the essential tools for measurement of results obtained from the business.
2. The capacity utilization of the selected sugar factories in relation to installed capacity is not satisfactory.
3. Sugar factories have not streamlined the total number of working days in the state.
4. To measure the total cost of processing of sugar cane, sugar factories must be correlated to the fixed cost with the variable cost.
5. Most of the sugar factories are running under loss because cost of production is more than the selling price of sugar and ineffective administration.
6. To evaluate the administrative efficiency of the sugar factories SWOP (T) management technique must be applied.

Objectives

Keeping all the points in view; an overall purpose of evaluating the performance of sugar industry in Karnataka has to specifically identify the following objectives:

1. To evaluate the pattern of ownership and performances of sugar industries in Karnataka.
2. To evaluate the growth and performance of selected sugar factories in the state in terms of both physical and financial indicators.
3. To analyze the business performance of the sugar factories under different ownership in Karnataka.
4. To determine and compare the capacity utilization and to arrive at the break-even level of production of

sugar in these factories both individually (based on ownership) and collectively.

5. To estimate the cost of processing of sugarcane into sugar, under different ownerships of sugar factories.
6. To identify the factors influencing the performance of sugar factories and evaluate the direction and magnitude of their influence
7. To identify the constraints in the functioning of the sugar factories and to suggest the appropriate strategies for improving the performance of sugar industry in the state.

Karnataka sugar industry incurred a loss of Rs 480 crores for the year 2007. The industry has urged both the state and central governments to grant few concessions, including waiver of purchase tax, deferment of value added tax implementation for next three years to tide over the crisis. Unless the government helps the industry to overcome the crisis, the industry will not be in a position to disburse the cane arrears. Cane arrears in the state for the year 2006 stands at Rs 498 crores, while for previous year it is about Rs 60 crores. The sugar industry in Karnataka is facing crisis due to over production of sugar for the last few years, which has accumulated into huge carry over stocks. Moreover, the prices are ruling low. While the cost of production is Rs 1,370 per quintal in Karnataka, the realisation is about Rs 1,070 forcing us to undergo a loss of nearly Rs 300 per quintal. Export of sugar is also not encouraging as the realisation is much lower at Rs 1,000 per quintal. Therefore the state government was urged to waive purchase tax on sugarcane for next few years and grant incentive of Rs 1,000 per tonne on exports as given by neighboring state Maharashtra and levy zero per cent VAT on sugar for next few years. The central government was requested by sugar industry to allow sugar mills to dispense with excise duty on sugar and molasses for a period of three years, convert the irregularity in cash credit into medium term loan on soft terms, reschedule the existing term loans into long-term loans at reduced interest rates. It was demanded that the government of Karnataka has to declare sugar industry as priority sector as it would enable the banking sector to extend more credit to the industry.

Problems of sugar industry in Karnataka: Two types of problems are identified in sugar industry in Karnataka. They are problems in long period and short period.

Short run problems: The important short run problems indented here are:

1. Fall in the sugar price - lower than the cost of production resulting in poor realization on carry over stocks.
2. Non-lifting of levy sugar- higher than the market price and high cost of holding sugar.
3. Reduced productivity and sugar recovery.
4. Drying up of working capital.
5. Shortage of cane and non- availability of pre-operative loans to factories.

Long run problems:

1. Lack of professional management, particularly in Co-operative and Public Sector Undertakings and excess staff
2. Avoidable administrative expenditure and over inventory
3. Fixing cane price above Statutory Minimum Price.
4. Many factories with less than 2500 TCD and unhealthy competition for raw martial
5. Obsolete technology and need for modernization.
6. Production of Ethanol, co-generation and chemicals necessary for optimal returns.

Policy prescriptions: Crucial policy prescriptions for the overall development of sugar industry in Karnataka are;

1. Extend Credit facilities to farmers and factories.

2. Conversion of overdue crop loans (principal & interest) to medium term loans.
3. Liberalization of refinance facilities through co-operative banks, commercial banks and RRB's to enable fresh borrowings by farmers.
4. Conversion of working capital overdues and short margins of sugar factories into medium term loans with moratorium of three years.
5. Rephase overdue longterm and other loans of the sugar factories.
6. Provide sugar factories loans for pre-operative expenses – without insisting on security and payable within six months.
7. Provide Working capital to factories against storage of sugar by relaxing exposure norms.
8. Levels of margin money on pledge loans availed by sugar factories should not be more than 10 per cent on levy sugar and 15 per cent on free sale sugar.
9. Settlement of all outstanding claims of factories by the government of India.
10. Ban of import of sugar and allow importing of raw sugar for reprocessing and export only.

CHAPTER 4

FINANCIAL PERFORMANCE OF SUGAR INDUSTRY IN KARNATAKA: CONCEPTS AND CONTEXT

This chapter deals with the description of the study area, the sampling techniques adopted, the nature and sources of data, the method of data collection and the various tools and techniques employed in analysing the data and evaluating the results. Apart from these afore said aspects, it also contains the precise definitions and explanations of some of the most important concepts and terms used in the study to facilitate clear understanding of the issues with which the present study is concerned. The present chapter is presented under the following major headings:

4.1 Description of the study area
4.2 Sampling design adopted
4.3 Nature and sources of data
4.4 Identification of performance indicators
4.5 Analytical techniques employed

4.1 Description of the Study

4.1.1 *Study region*

The present study was conducted in the State of Karnataka. Karnataka is the eighth largest state in India, both in terms of its area and population. It has a total geographical area of 1, 19,257 Sq. Kms (5.84 per cent of India's total area) and a population of about 53.34 million (5.31 per cent of India's population), with a density of 256 persons per square kilometer (2001 census).

Karnataka is situated in the west central part of peninsular India. It consists of a narrow elongated belt between the Arabian sea and the western *ghats* and enhanced coastline of about 400 kilometres. The State is situated between 11^0 and 19^0 North latitude and 74^0 and 78^0 East longitudes. The State is bounded by Maharashtra on the north and by Goa and the Arabian Sea on the west. It has a common border with Andhra Pradesh on east and with Tamil Nadu and Kerala on the south.

The climate of the state may be described as essentially tropical monsoon type, which is a product of the interplay of the two opposing in masses of the southwest and northeast monsoons. Over greater parts of the State, summers are languorously warm and winters are bracingly cool. In the coastal region, humidity makes summer enervating. In the northern region, winters are delightfully cool but summers are uncomfortably hot. The major portion of the rainfall is from the south west monsoons. There are regional variations in the distribution of annual rainfall and it varies from 7620 mm on the western-*ghats* to about 381 mm. in the eastern and the north eastern parts. The rainfall ranges from 559 mm. in Chitradurga and Bijapur districts to about 3860 mm. in Dakshina Kannada district.

Regional variations are found both in the temperature and the distribution of annual total rainfall. The coastal region has mean monthly temperatures of 24^0 to 30^0 C all the year round. In Malnad, the temperature rhythms are very similar to those of the coastal region. But the region is relatively cooler because of the higher altitudes. Here the mean monthly temperature varies between 18^0C and 24^0C. Malnad is characteristically a region of heavy rainfall with a single peak in July. The northern *maidan* region is of hot summers; warm winters and of rather sparse and variable rainfall on the whole. The mean monthly temperature varies between 22^{0}C and 25^{0}C. Rainfall mainly comes in the form of torrential storms during the period of southwest monsoons usually with a September peak. The south *maidan* region rather cool with equable temperature. The annual rainfall is between 508 mm. to 762 mm. The drier areas in the north eastern part around Chitradurga and in the extreme east near Kolar and the variability of rainfall are also high in these areas. The average rainfall for the year is about 1200 mm. It varies from 7,620 mm. to 380 mm. The normal rainfall for the state is 1,315 mm.

4.1.2 *Agricultural zones of Karnataka*

Based on rainfall pattern, topography, soil characteristics, cropping patterns etc. Karnataka has been grouped into ten Agro-Climate Zones.

1. North Eastern Transition Zone, 2. North Eastern Dry Zone, 3. Northern Dry Zone, 4. Central Dry Zone, 5. Eastern Dry Zone, 6. Southern Dry Zone, 7. Southern Transition Zone, 8. Northern Transition Zone, 9. Hilly Zone and 10. Coastal Zone.

4.2. Sampling Design Adopted

For studying the growth and performances of sugar industry in Karnataka as a whole, all the sugar factories working in the State were considered. Among all the sugar factories sampling is formed on the basis of specific criteria to collect the required data

4.2.1 *Selection of Sugar Factories*

By 2007, in Karnataka, there were 52 sugar factories functioning. Of these 18 were functioning under co-operative sector, (forming 10 per cent of the total) followed by 32 factories under private sector (88 per cent of the total) and two factories under public sector (two per cent of the total).

The sampling method for the study used is convenient sampling, which is of non-probabilistic nature and is basically adopted on the basis of specific criteria. The criteria for selection of the sugar factories are ownership, availability of data, profitability position and financial strength of the sugar factories working in Karnataka.

The selection of sugar factories was done in three phases; in the first phase the sugar factories were classified into three groups, based on the ownership. These groups were as follows:

1. Sugar factories in Co-operative sector
2. Sugar factories in Public sector and
3. Sugar factories in Private sector

In the second phase, these sugar factories in each sector were classified as above average (profit making sugar factories) and below average factories (loss incurring sugar factories) with respect to the following physical parameters.

(a) Installed capacity during 2004-05,
(b) Sugarcane crushed during 2004-05,
(c) Sugar produced during 2004-05 and
(d) Recovery percentage of sugar from sugarcane during 2004-05

As there were 18 sugar factories in co-operative sector, 32 under private ownership and two under public sector functioning in Karnataka. Out of this, six sugar factories under co-operative sector (three above average and three below average), and two sugar factories each (one above average and one below average) working under public and private sector were selected for the study.

Thus, the names of the sugar factories so selected were as under:

A. Co-operative Sector

1. Above average Sugar factories
   (i) Malaprabha Co-operative Sugar Factory Limited, M.K.Hubli

(ii) Hiranyakeshi Co-operative Sugar Factory Limited, Sankeshwar

(iii) Dood Ganga Krishna Co-operative Sugar Factory Limited, Chikodi

2. Below average sugar factories

(i) Pandavapura Co-operative Sugar Factory Limited, Pandavapura

(ii) Hemavathi Co-operative Sugar Factory, Srinivaspura

(Iii) Ghataprabha Co-operative Sugar Factory, Gokak

B. Private Sector

1. Above average sugar factory
   Ugar Sugar Works, Ugarkhed
2. Below Average Sugar Factory
   Davanagere Sugar Mills, Kukkuwadi

C. Public Sector

1. Above average sugar factory
   My Sugar Company Limited, Mandya
2. Below average sugar factory
   Mysore Paper Mills (Sugar Division), Bhadravathi

4.3 Nature and Sources of Data

The data were collected from various sources to evaluate the objectives of the study.

4.3.1 For evaluating the growth and performance of sugar industry in Karnataka as a whole, the data on number of sugar factories, installed capacity, cane crushed, sugar production, recovery percentage of sugar and duration of season were collected factory-wise for a period of 15 years from 1990-91 to 2004-05 from the following sources:

(a) Karnataka State Co-operative Sugar Factories Federation and South India Sugar Mills Association.

(b) South India Sugar Mills Association (Karnataka Branch)
(c) Indian Sugar Mills Association, New Delhi
(d) Commissioner of Cane Development, Directorate of Sugar, Government of Karnataka, Bangalore
(e) National Federation of Co-operative Sugar Factories Ltd., New Delhi
(f) Karnataka Brewers and Distillers Association, Bangalore
(g) Sugar Technologists Association of India, New Delhi
(h) All India Distillers Association, New Delhi
(i) India Sugar Export and Import Corporation Ltd., New Delhi
(j) Directorate of Sugar, Ministry of Consumer Affairs, Food and Public Distribution, New Delhi.

4.3.2 For evaluating the growth and business performance of the selected sugar factories, the data relating to physical and Financial aspects were collected for a period of 15 years from 1990-91 to 2004-05 from the following sources:

(a) Yearly summary statement of individual sugar factories
(b) Balance sheet
(c) Profit and loss account
(d) Production and manufacturing account
(e) Receipts and payments account and trading accounts
(f) Schedule of Accounts and Investments
(g) Statement of financial position and working results
(h) Quantity details account
(i) Cash flow and funds flow statements
(j) Cost audit report, of each selected sugar factory were abstracted from the annual reports and audit reports of the respective sugar factories for a period of 15 years from 1990-91 to 2004-05.

4.3.3 For evaluating the cost of processing and to arrive at the break even level of processing of the factories, individually (based on ownership) and collectively, the relevant data were collected from the annual production records and annual financial records of the individual Sugar Factories and analysed for 15 years from 1990-91 to 2004-05 except for the one below average co-operative sugar factory i.e., Pandavapura Co-operative Sugar Factory, since the data were not available for 2003-04 and 2004-05 (The factory closed for these two years).

4.4 Identification of Performance Variables

Twenty two performance indicators were identified to have close association with the functioning of the factories. Of them 8 indicators related to physical aspects and they were termed as physical indicators, while the remaining 24 indicators related to the financial indicators.

Physical and financial indicators selected for the study:

I. Physical indicators/variables

1. Installed Capacity, 2. Sugar Production, 3. Cane Crushed, 4. Cane Area in District, 5. Cane Yield in District, 6. Recovery Percentage, 7. Duration of Crushing and 8. Capacity Utilisation.

II.Financial Indicators/Variables

1. Share Capital, 2. Reserve Funds, 3. Equity, 4. Net Profits, 5. Owned Funds, 6. Borrowings, 7. Current Liabilities, 8. Total Liabilities, 9.Total Inventory, 10. Sugar Inventory, 11. Current Assets, 12. Net Working Capital, 13. Fixed Assets, 14.Total Assets, 15. Opening Stock (sugar), 16. Opening Stock (Total), 17. Average Inventory, 18. Total Sales, 19. Sugar Sales, 20. Sale of By-products, 21.Total Income, 22.Total Purchases, 23.Total Expenses and 24. Net Worth

4.5. Analytical Techniques Employed

The analyses was carried out in twelve different phases.

The first phase included the analyses of basic information of the sugar industry in the State as a whole, relating to physical indicators and computation of their growth rates. The second phase was devoted to the analyses of basic information of the selected sugar factories, relating to both physical and financial indicators. In the third phase, growth rates were calculated for important physical and financial performance indicators of the sample sugar factories, both individually (based on ownership) and collectively. The financial ratio analyses was carried out in the fourth phase, both individual factory-wise and collectively. The fifth phase, the performance of the sugar factories was analysed by employing Dupont Analyses Technique. The sixth, seventh and eight phases of analyses were devoted respectively to analyze the break-even analyses, safety margin, capacity utilisation, and cost of processing, both factory-wise and collectively. The ninth, tenth, and eleventh phases are devoted for analysing the data by using statistical tools like, Standard Error estimate of Standard Error by using least square method and t-Test. The twelveth phase of analyses included analyses of an overall view of the management of the´ sugar industry employing SWOP (T) management technique.

4.5.1 *Tabular presentation*

Tabular presentation was employed to study the performance of sugar industry in Karnataka as a whole in terms of physical indicators as well as financial indicator to evaluate the performance of selected sugar factories. Simple statistical tools/measures like averages, Standard Error, Variances and t-Test percentages and ratios have been used in the study.

4.5.2 *Ratio analyses*

Ratios are the measurements of results obtained from a business. In order to know the performance of the selected sugar factories in the State, the financial ratio analyses, an effective tool providing the clear understanding of the performance of a business organisation, has been heavily relied upon. The ratios relating to solvency, liquidity, profitability, turnover, efficiency and strength of these factories have been analysed. To highlight each of the above said aspect, different ratios under each category have been worked out.

The secondary data drawn from the audited annual statements of the balance sheet, reports of receipts and payments account, profit and loss account of the selected sugar factories for a period of fifteen years from 1990-91 to 2004-05. The financial ratios used for the analyses are described below.

4.5.2.1 *Tests of Liquidity*

For continuous operation of a business organisation, its liquidity would be basic. The liquidity ratios, also called as balance sheet ratios, are used to measure the ability of an enterprise to possess adequate cash to meet the immediate cash obligations. The degree of liquidity of the sugar factories, both individually and collectively, was determined employing the following ratios.

(a) Current Ratio

$$\textbf{Current Ratio} = \frac{\textbf{Current Assets}}{\textbf{Current Liabilities}}$$

The Current ratio would give the relationship between current assets and current liabilities of the sugar factories.

(b) Acid Test Ratio

This ratio, also called as quick ratio or near money ratio, represents the ratio between quick assets and current liabilities, and was computed by using the following formula.

$$\textbf{Acid Test Ratio} = \frac{\textbf{Current assets - Inventory}}{\textbf{Current liabilities}}$$

The ratio would indicate the dependence of an organisation on inventory for liquidity.

(c) Liquid Assets to Total Assets Ratio

$$\textbf{Liquid Assets to Total Assets Ratio} = \frac{\textbf{Liquid assets}}{\textbf{Total assets}}$$

Liquid assets includes cash on hand and in bank, inventory, bills and accounts receivable and deposits in banks, which could be readily withdrawn on demand.

(d) Inventory Ratio

$$\textbf{Inventory Ratio} = \frac{\textbf{Inventory}}{\textbf{Net working capital}}$$

Where,

Net working capital = Current assets – current liabilities

4.5.2.2 *Tests of Solvency*

The solvency ratio would reflect on the medium term and long-term solvency position of the sugar factories, indicating its ability to meet the medium term and long-term obligations and these ratios would provide basis for measuring the leverage effect on the performance of the sugar factories. The solvency position of the sugar factories, both individually and collectively, was computed by employing the following ratios.

(a) Total Liabilities to Owned Funds Ratio

This ratio is obtained as given below.

$$\textbf{Total Liabilities to Owned Funds Ratio} = \frac{\textbf{Total liabilities}}{\textbf{Owned funds}}$$

(b) Fixed Assets to Owned Funds Ratio

This ratio is computed by dividing depreciated fixed assets by the owned funds of the sugar factory.

$$\textbf{i.e., Fixed Assets to Owned Funds Ratio} = \frac{\textbf{Depreciated value of fixed assets}}{\textbf{Owned funds}}$$

(c) Debt-Equity Ratio

$$\text{Debt-Equity Ratio} = \frac{\text{Long-term borrowings}}{\text{Share capital + Reserves}}$$

4.5.2.3 *Tests of Turnover*

The turnover ratios indicate the effectiveness with which different assets were utilised in the business. Turnover means the number of times an asset flows through a firm's operations into its total sales.

The important turnover ratios considered in the present study are described below.

(a) Inventory Turnover Ratio

This ratio measures the effectiveness of the sugar factory in its sales efforts. Higher the turnover ratio better would be the performance.

$$\text{Inventory Turnover Ratio} = \frac{\text{Annual sales}}{\text{Average inventory}}$$

Where,

$$\text{Average inventory} = \frac{\text{Opening stock + Closing stock}}{2}$$

(b) Total Assets Turnover Ratio

Total assets turnover ratio is computed by dividing total sales by the total assets of the factory in a given period of time.

$$\text{i.e. Total Assets Turnover Ratio} = \frac{\text{Total Sales}}{\text{Total assets}}$$

(c) Working Capital Turnover Ratio

$$\text{Working Capital Turnover Ratio} = \frac{\text{Total sales}}{\text{Total working capital}}$$

4.5.2.4 *Tests of Efficiency and Profitability*

The tests of efficiency indicate the utilisation of capital and income whereas the profitability tests are meaningful measures, which can be used to diagnose the financial status and the overall efficiency of a business organisation.

The efficiency and profitability of the sugar factories were measured by employing the following ratios.

(a) Efficiency of Capital Ratio

This ratio is used to evaluate the efficiency of sugar factories in regard to their capital utilisation.

$$\textbf{Efficiency of Capital Ratio} = \frac{\textbf{Paid up share capital}}{\textbf{Annual sales value}}$$

(b) Gross Ratio (percentage)

This ratio is computed as given below,

$$\textbf{Gross Ratio} = \frac{\textbf{Total Expenses}}{\textbf{Gross income}} \times 100$$

This ratio would help to ascertain how efficiently the gross income of the sugar factory was utilised.

3.5.2.5 *Tests of Financial Strength*

These ratios are employed to assess the real worth of the sugar factory, and are described below.

(a) Net Worth

Net Worth = Total assets – Total liabilities

Net worth measures the excess of assets over the liabilities. Wider the positive difference, sounder would be the net worth of sugar factory.

(b) Net Capital Ratio

This ratio would reflect the financial strength of a business organisation.

$$\text{Net Capital Ratio} = \frac{\text{Total assets}}{\text{Total liabilities}}$$

Higher the ratio, greater would be the margin of safety against the decline in the prices of major assets of the sugar factory.

(c) Equity Capital Ratio

This ratio would reflect the composition of owner's equity and is worked out as follows.

$$\text{Equity Capital Ratio} = \frac{\text{Owner's equity}}{\text{Paid-up share capital}}$$

4.5.2.6 *Fixed Assets Ratios*

These ratios act supplementary to the earlier ratios. The important fixed assets ratios considered in the present study are described below.

(a) Tests of Profitability

The data regarding the total expenses and gross income were drawn from the profit and loss account of the sugar factory over a period of time. The gross income after deleting net profit as shown in the profit and loss account is considered.

(b) Net profits to Owned Funds Ratio

This ratio reveals the extent of profitability in relation to investment of owned funds of the sugar factory and is compounded as under

$$\text{Net profits to Owned Funds Ratio} = \frac{\text{Net Profits}}{\text{Owned funds}}$$

(c) Net Profits to Total Assets Ratio

This ratio represents the earning capacity of the total assets of the sugar factory over a given period of time.

(d) Net Profits to Total Sales Ratio

Net profits to total sales ratio also known as net profit margin, indicates a firm's ability to earn profits after meeting all direct and indirect costs of business. This ratio is computed by dividing net profits by total sales.

$$\textbf{i.e. Net Profits to Total Sales Ratio} = \frac{\textbf{Net Profits}}{\textbf{Total sales}}$$

(e) Fixed Assets to Total Assets Ratio

This ratio is worked out by using the following formula.

$$\textbf{Fixed Assets to Total Assets Ratio} = \frac{\textbf{Fixed assets}}{\textbf{Total assets}}$$

This ratio would reflect the proportion of fixed assets in the total assets of the sugar factory. Higher ratio depicts that a higher percentage of the assets are fixed assets.

(f) Fixed Assets to Net worth Ratio

This ratio is computed by dividing fixed assets by the net worth of the factory at a given point of time.

$$\textbf{i.e. Fixed Assets to Net worth Ratio} = \frac{\textbf{Fixed assets}}{\textbf{Net worth}}$$

4.5.3 ***Dupont Analyses***

The Dupont Formula shows the interrelationship between five key financial ratios. It can be presented in several ways.

Profit Margin	×	Total Asset Turnover	=	Return on Investment				
				Return on Investment	×	Equity Multiplier	=	Return on Equity
Profit Margin	×	Total Asset Turnover			×	Equity Multiplier	=	Return on Equity

Each of the key ratios will now be defined in turn.

Profit Margin (PM) = Net Income ÷ Sales (or Total Revenue)

Total Asset Turnover (TAT) = Sales (or Total Revenue) ÷ Total Assets

Return on Investment (ROI) = PM × TAT= Net Income ÷ Total Assets

Equity Multiplier (EM) = Total Assets ÷ Total Equity

Return on Equity (ROE) = PM × TAT × EM

= ROI × EM

= Net Income ÷ Total Equity

4.5.4 ***Break-Even Analyses (BEA)***

$$\textbf{BEP (Units)} = \frac{\textbf{TFC}}{\textbf{P} - \textbf{V}} = \text{“in Units”} \ldots\ldots\ldots\ldots (1)$$

To analyze the break-even level of sugar production, the following methodology is adopted.

Let TR = Sales or Revenue (Rs)

P = Selling Price of Sugar (Rs/Qtl)

TC = Total Cost of Processing Sugar (TFC+TVC)

TFC = Total Fixed Cost (Rs)

TVC=Total Variable Cost (Rs/Qtl of Sugar Production)

N = Total Sugar Production (Qtls)

AVC = TVC / N (Rs/Qtl of Sugar Production)

BEP = Break-Even Level of Sugar Production (Qtl)

Profit =TR – TC

Where, TR = NxP, and

TC = TFC + TVC

Hence, Profit = TR-TC

At BEP , Profit/Loss = 0

Solving equation (1), the Break-Even level of sugar production was calculated for each sugar factory, individually, for the year 1990-91 to 2004-05.

4.5.5 *Safety Margin*

It is crucial and used as management control technique to determine optimum sales, which yields safety margin to a firm. Safety margin is expressed in percentage:

The formula used to calculate the safety margin of individual sugar factories is expressed as below:

$$\mathbf{SM} = \frac{\mathbf{Sales - BEP(units)}}{\mathbf{Sales}} \mathbf{X\ 100}$$

While calculating the Safety Margin, only sugar sales are considered, not the total sales because the sugar production are predominant.

Some of the assumptions, which hold good for the study in calculating safety margin, are:

1. The total cost is divided into fixed cost and variable costs of sugar production.
2. The selling price is constant
3. The average and marginal productivity of factors are constant.

4.5.6 *Standard Error*

The standard error of a statistic is the standard deviation of the sampling distribution of that statistic. Standard errors are important because they reflect how much sampling fluctuation a statistic will show. Here, standard error quantifies the uncertainty that comes from measuring only a sample sugar factories rather than measuring the entire sugar industry.

For the present research we have calculated the standard error by using the formula

$$\text{SE} = \sqrt{\frac{\Sigma(\text{X-Mean})^2}{\text{N-1}}}$$

Where X = Indicators' Value

N= Sample Size

Selected physical and financial indicators used to determine the standard error

Physical Indicators

a) Sugar Production, b) Quantity of Sugarcane Crushed, c) Recovery Percentage of sugar and d) Capacity Utilisation

Financial Indicators

a) Share Capital, b) Net Profit, c) Barrowings, d) Current Liabilities, e) Level of reserve funds, f) Total expenses, g) Gross income, h) Total assets position, i) Total sales and j) Fixed assets position

Procedure:

Given Sugar Production as one Physical Indicator or Variable (X) of Selected sugar factories.

1. Calculation of the Mean Value of 1990-91.
2. Deducted Mean Value from Individual Sample Value i.e. (X-Mean)
3. Squaring of the (X-Mean) Value
4. Obtaining the Sum of Square of (X-Mean)
5. Dividing the step 5 by (N-1) Where N is Sample size.
6. Taking Square Root of Step 6 obtains Standard Error.

4.5.7 *Estimate Of Standard Error Based On Least Square Technique*

The main objective of many statistical investigation is to make predictions preferably on the basis of mathematical equation. Equivalently we write the equation as:

$Y_i = \alpha + \beta_i + e_i$ for i = 1,2,……n.

By solving the above equation the standard error of estimate is calculated by using the residual sum of squares divided by (n-2). This is denoted as below:

Estimate of $\sigma^2 = S^2 =$

$$S^2_e = \frac{Syy - (Sxy)^2/Sxx}{n-2}$$

Alternatively we have used the formula to calculate the standard error of estimate.

$$S_{yx} = \sqrt{\frac{1}{n}\Sigma(y-y_e)^2}$$

For estimation of standard error also, 14 selected parameters are used as mentioned above in the standard error calculation.

4.5.8 *t-Test*

Here, sample size is less than 30 each year's observation values are considered as a sample elements and developed both null hypothetical statement(H_0) as well as alternative hypothetical statement(H_A) where in alternative hypothesis is subjected to verify. It is calculated that t-value assuming the level of significance at 5 per cent with the degree of freedom (n-1) assuming to be normal distribution. The formula used to calculate t-value is as below:

Formula

$$t = \frac{\overline{x} - \mu}{\sqrt{S^2/n}}$$

Where

$$\overline{x} = \frac{\Sigma x}{n}$$

$$S^2 = \frac{1}{n-1}\Sigma(x-x)^2$$

S = Sample Standard Deviation

μ = Assumed Sample Mean

4.5.9 *Compound Growth Rate (CGR)*

Least-squares growth rates are used whenever there is a sufficiently long time series to permit a reliable calculation. No growth rate is calculated if more than half of the observations in a period are missing. The least-squares growth rate, *r*, is estimated by fitting a linear regression trend line to the logarithmic annual values of the variable in the relevant period. The regression equation takes the form

$\ln X_t = a + bt,$

This is equivalent to the logarithmic transformation of the compound growth equation,

$X_t = X_o (1 + r)^t = X_o (1 + r)^t = X_o (1 + r)^t$

In this equation X is the variable, *t* is time, and *a* = ln Xo and *b* =

These (1 + r) are parameters to be estimated. If *b** is the least-squares estimate of *b*, the average annual growth rate, r, is obtained as [exp(b^0) – 1] and is multiplied by 100 for expression as a percentage.

4.5.10 *Estimation of Capacity Utilisation*

The capacity utilisation is computed using the following formula.

$$\text{Capacity Utilisation (in \%)} = \frac{\text{Cane Crushed During the Year}}{\text{Installed Crushing Capacity/Day x Total Number of Working Days}} \times 100$$

In the above formula, total number of working days of each sugar factory are considered for a period of one year (2004-05).

4.5.11 *Cost of Processing*

The total cost of processing sugarcane into sugar is made up of fixed costs and variable costs. The fixed cost included salaries and wages, welfare expenses, provision for bonus, auditors expenses, TA and sitting charges of directors, insurance premium paid, postage charges, printing and stationary expenses, meeting expenses, miscellaneous expenses, maintenance of buildings, vehicles, rent and taxes, expenses on sugar distribution, interest paid and depreciation of plant and machinery.

On the other hand the variable cost included cost of cane purchased, transportation cost, harvesting charges, purchase tax, cost on cane development activities and loading and unloading changes of cane consumable stores, electricity charges, maintenance of plant and machinery, sugar bagging and stocking costs and freight and transportation charges.

The cost of processing was computed for all the ten sugar factories individually and collectively for the accounting year 2004-05.

4.5.12 *SWOP (T) Management Technique*

It should pay more attention to its weakness. Prompt actions should be taken to remedy the weaknesses thereby putting the institution on proper position. It should explore the new opportunities in the offing so that it can seize it initial advantages and stabilize its position in new areas before the competition is intensified in the interest of overall growth and stability of the institution.

Keeping these things in view, the SWOP (T) management approach has been employed to identify specific strengths (S) as well as weaknesses (W) of the sugar factories in their operation and identify new opportunities (O) which would help in combating the unforeseen problem (P) or threat (T) that may crop up.

CHAPTER 5

ANALYSES OF FINANCIAL PERFORMANCE OF SUGAR INDUSTRY IN KARNATAKA

The main focus of this chapter is to throw light on some of the causes responsible for the major trends observed in financial performance of sugar industry in Karnataka. The analyses of the major trends in terms of causes as well as their interaction are made. It is hoped that, it would help in identifying some policy measures, which could be adopted to overcome the constraints the sugar factories are suffering.

The analyses of the financial performance of sugar industry during the study period was evaluated in by using both financial tools and statistical tools. Financial tools were used to evaluate the financial strength of the sugar factories and statistical tools were used to know the accuracy of data used in the study. The important financial tools used in the study are presented here:

5.1 Pattern of ownership and performance of sugar industry in Karnataka.

5.2 Performance and growth of sugar factories.

5.3 Financial analyses of sugar factories.

5.4 Statistical analyses of the sugar factories.

5.5 Capacity utilisation of sugar factories.

5.6 Cost of processing sugarcane into sugar.

5.1 Pattern of Ownership and Performance of Sugar Industry In Karnataka

In the performance evaluation of any organisation, the study of the basic information is indispensable, since, it provides clear and deeper insight into the basic structure of the organisation, its growth and development. The basic data pertaining to sugar industry, as a whole is colleted for the period of 15 years from 1990-91 to 2004-05, the results are presented under the following sub-heads.

As already described in methodology, at present there are 35 sugar factories functioning in the state, of which 15 are in co-operative sector, 18 and two are under private and public sector, respectively.

5.1.1 *Sugar production in Karnataka*

The sector-wise and year-wise sugar production in the state is presented in the Appendix-1.

The total sugar production in the state increased from 9,38,983 tonnes in 1990-91 to 15,46,453 tonnes during 2004-05 with an average growth rate of 4.34 per cent.

The contribution of private sector ranged between 37.84 per cent to 64.78 per cent of the total sugar production in the state, followed by co-operative sector (ranging between 48.12 per cent and 33.93 per cent) and Public Sector (14.04 per cent to 1.25 per cent). The average growth rate of sugar production by factories under private sector, and public sectors were 4.34, 1.58 and –4.06 per cent per anunm, respectively.

There are a lot of fluctuations in the production of sugar in co-operative and private sector. During 1990-91 there was more production incase of public sector, where as during 2004-05 the production decreased to 19,270 tonnes, this is because of changing government policies. The compound growth rate of public sector shows the negative value (–4.06 per cent) because of fluctuations in the sugar production.

5.1.2 *Sugarcane crushed in Karnataka*

The quantities of sugarcane crushed, sector-wise and year-wise over the period from 1990-91to 2004-05 are represented in Appendix-2.

In Karnataka the sugarcane crushed by the sugar factories for the production of sugar increased from 91,65,889 tonnes during 1990-1991 to 6,98,31,340 tonnes during 2004-05 with an average growth rate of 11.98 per cent. On an average, 46.46 per cent of total cane was crushed by co-operative sector sugar factories, followed by private sector (42.25 per cent) and public sector (11.30 per cent) sugar factories. The annual average growth rates of cane crushed were 16.31, 8.60 and –1.88 per cent by the sugar factories under co-operative, private and public sector, respectively. There is a negative growth rate of sugarcane crushed (–1.88 per cent) due to internal problems of the public sector sugar factories.

5.1.3 *Average duration of crushing*

The average crushing season (days) of the sugar factories functioning in the state is presented in Appendix-3.

When sugar industry was considered, in the state as a whole the average season ranged between 120 days during 1997-98 and 199 days during 2004-05. The crushing duration was found to be low during 1997-98, 1993-94 and 2003-04 and gradually increased over the years. The average duration of crushing by the factories under private, co-operative, and public sector was 158,155 and 152 days, respectively. The crushing duration ranged between 135 and 201 days in private sugar factories; between 110 days and 213 days in co-operative sector sugar factories and between 132 days and 183 days in public sector sugar factories during the period.

5.1.4 *Recovery percentage of sugar*

The recovery percentage of sugar from cane over a period of 1990-91 to 2004-05 Sector-wise and year-wise is presented in Appendix-4. On an average the sugar recovery was 10.46

per cent in the state over the period of study. The average recovery of sugar from sugarcane was 10.63 per cent in co-operative sugar factories followed by 10.52 and 10.18 per cent in the sugar factories under private and public sectors respectively. The year-to-year fluctuations in the recovery percentage of sugar were minimum.

5.2 Performance and growth of sugar factories

The basic information pertaining to both physical and financial indicators of ten selected sugar factories from 1990-91 to 2004-05 were collected and analysed. The information was further stratified into different strata based on the ownership of the sugar factories. The physical and financial indicators selected reflect the structure and performance of the factories for further analyses.

5.2.1 *Physical indicators*

The importance of physical indicators considered for the study were installed capacity, sugar production, cane crushed, recovery percentage of sugar, duration of crushing season and utilisation of the installed capacity. The performances of these physical indicators are described below under different sub heads.

5.2.1.1 *Installed capacity*

The installed capacity in tonnes of cane crushing per day (TCD) of different sugar factories over the period of study is presented in Appendix-5 and 5a. The overall average installed capacity increased from 3097.22 TCD during 1990-91 to 4,177 TCD during 2004-05, with an average growth rate of 2.93 per cent per annum. In co-operative sugar factories the installed capacity increased from 2416.67 TCD during 1990-91 to 3158 TCD during 2004-05 with a growth rate of 2.28 per cent. The installed capacity grew at the rate of 4.54 and zero per cent in private and public sector sugar factories respectively. Among the below average factories, a growth rate of 4.82 per cent per annum was observed in co-operative sector, whereas it was zero per cent in private sector, (Davanagere Sugar Factory) and installed capacity remained unchanged in public sector (Mysore Paper Mills, Bhadravathi) similarly among the above average co-operative sugar factories, the highest growth rate was found in the case of Doodhganga co-operative sugar factory (13.55 per cent) and installed capacity remain unchanged in other two above average co-operative sugar factories (Hiranyakeshi and Malaprabha co-operative sugar factories). In the above average public sector co-operative sugar factories (My sugar Mandya) the installed capacity remain unchanged.

5.2.1.2 *Sugarcane crushed*

Average quantity of sugarcane crushed by the sugar factories increased from 5,12,395 tonnes during 1990-91 to 6,12,868 tonnes during 2004-05, with a growth rate of 1.57 per cent (Appendix-6 and 6a).

The quantity of sugarcane crushed is highest in private sector i.e., 8,96,314 tonnes during 2004-05 with a growth of 3.25 per cent. In public sector sugar factory quantity of sugarcane crushed during 2004-05 was 6,05,327 tonnes with a growth rate of 1.14 per cent. During 2004-05, co-operative sugar factories crushed 3,36,963 tonnes with a negative growth rate of –0.72 per cent.

5.2.1.3 *Recovery percentage*

The factory-wise, year-wise, recovery percentage of sugar from sugarcane is presented in the Appendix-7 and 7a. The table revealed that on an average, the recovery percentage of sugar from sugarcane was 10.08 per cent over the study period.

The average recovery percentage was the highest in private sector sugar factories (10.72 per cent) followed by co-operative sugar factories (10.10 per cent) and public sector sugar factories (9.42 per cent). Among the below average factories, the average recovery was the highest in co-operative sector sugar factory (Gataprabha sugar factory 11.04 per cent) followed by private sugar factory (Davanagere sugar factory, 9.80 per cent) and public sector (Mysore paper Mills 10.07 per cent).

Private sugar factories performed well with respect to the recovery percentage (11.64 per cent) among the above average factories followed by co-operative and public sector sugar factories (11.49 per cent and 8.78 per cent, respectively). The highest recovery percentage of sugar from sugarcane was found in the case of Ugarkhurd sugar factory (12.74 per cent) during 1992-93 and lowest was in the case of My Sugar factory Mandya (5.36 per cent) during 2004-05.

5.2.1.4 *Duration of crushing season*

The duration of crushing season depends mainly on the availability of sugarcane in time and in required quantity.

Over the study period, the over all average duration of crushing season was 201 days (Appendix-8 and 8a). The average duration of crushing was 183, 219 and 200 days in co-operative sector sugar factories, public sugar factories and private sugar factories, respectively. The highest duration of crushing was found in the case of Mandya sugar factory (247 days) followed by Ugarkhurd sugar factories (213 days) and Hemavathi co-operative sugar factory (204 days). My Sugar and Ugarkhurd sugar factories are above average sugar factories and Hemavathi co-operative sugar factory is below average sugar factory.

5.2.1.5 *Sugar production*

Sugar produced by different sugar factories from 1990-91 to 2004-05 is presented in Appendix-9 and 9a.

The overall average sugar production increased from 5,38,769 quintals during 1990-91 to 5,67,552.11 quintals 2004-05, with an annual growth rate of 0.91 per cent per annum.

The average sugar production in co-operative sugar factories decreased from 4,43,716 quintals during 1990-91 to 3,03904.33 during 2004-05 with a negative growth of –0.78 per cent. Incase of public sector sugar factories also the production of sugar decreased from 491622 quintals during 1990-91 to 383620 quintals during 2004-05 with a negative growth rate of –2.06 per cent. Only incase of private sector average sugar production increased from 680968.50 quintals during 1990-91 to 1015132 quintals during 2004-05 with a positive growth rate of 2.99 per cent.

5.2.2 *Financial Indicators*

The financial management performance of sugar factories could be judged by examining the trend of various financial indicators such as share capital, owned funds, asset position, working capital, liabilities position, sales, expense, gross income and net profits. These financial indicators are described below for the study period from 1990-91 to 2004-05.

5.2.2.1 *Share capital*

Share capital position of different sugar factories over the years is presented in Appendix-10 and 10a.

The overall average share capital increased from Rs.508.26 lakhs during 1990-91 to Rs1464.36 lakh during 2004-05, with a growth rate of 7.29 per cent. The share capital in private factories grew at a rate of 15.05 per cent per annum. It increased from Rs. 198.99 lakhs in 1990-91 to Rs.1772.77 lakhs in 2004-05. In co-operative sugar factories, share capital grew at the rate of 5.60 per cent per annum, whereas the growth rate of share capital was just 5.05 per cent per annum in public sector sugar factories. The rate of growth of share capital was 19.50 per cent in private sector sugar factories (from Rs. 201.99 lakhs during 1990-91 to Rs.2983.04 lakhs during 2004-05, Davanagere sugar factory), followed by co-operatives (12.92 per cent) and public sector factories (10.13 per cent) among the above average sugar factories. However, a steady growth of share capital was found in the case of below average co-operative (3.15 per cent) and public sector (5.05 per cent) sugar factories. The growth rate was found to be negative in the case of below average co-operative sugar factories (Pandavapura S.F –2.68 per cent).

5.2.2.2 *Reserve funds*

The overall average reserve funds increased from Rs.545.87 lakhs in 1990-91 to Rs.998.04 lakhs in 2004=05, at the growth rate of 5.71 per cent per annum (Appendix-11 and 11a).

Among the different sectors, the private sugar factories had maintained high level of reserve funds compared to public and co-operative sector sugar factories. The private sugar factories accumulated the reserve funds at an average annual growth rate of 15.59 per cent. In the case of public sector sugar factories, the reserve fund was Rs.1115.07 lakhs in 1990-91 to Rs 363.95 lakhs in 2004-05 with a negative growth rate of – 4.73 per cent.

The rate of reserve fund accumulation was very high in the case of Ugar sugar factory (above average private), as compared to (8.95 per cent) above average co-operative and public sector (-4.73 per cent) But the reserve funds decreased from Rs.1625.31 lakh in 1990-91 to Rs.200.90 lakhs in the case of below average public sector sugar factory (Mysore Paper Mills, Bhadravathi).

5.2.2.3 *Owned funds*

The total funds owned by the sugar factories are presented in Appendix-12 and 12a.

The overall average amount of funds a factory owned increased from Rs. 1054.10 lakhs in 1990-91 to Rs. 2462.40 lakhs during 2004-05 with a growth rate of 6.95 per cent per annum.

The level of owned funds was higher in the case of private sugar factories compared to public and co-operative sugar factories, and the rate of growth of owned funds also higher in private sugar factories (15.40 per cent).

Among the below average sugar factories, private sector (Davanagere) ranked first in growth rate (19.72 per cent), whereas in the case of co-operative the amount increased from Rs. 976.48 lakhs to Rs. 1650.08 lakhs during 1990-91 to 2004-05 period.

In the case of above average sugar factories, a steady growth of owned funds was observed in the case of co-operatives (5.71 per cent), ranking first with respect to the amount and second with respect to the growth rate, first being private sector (Davanagere Sugar Factory, 19.72 per cent). The rate of accumulation of owned funds was comparatively low (4.71 per cent) in public sector sugar factory.

5.2.2.4 *Assets position*

Overall average current assets (Appendix-13 and 13a), fixed assets (Appendix-14 and 14a) and the total assets (Appendix-15 and 15a) grew at the annual growth rate of 8.55, 12.82 and –11.07 per cent, respectively.

When sector-wise breakup was considered, the current assets increased at the rate of 9.65, 6.31 and 9.39 per cent per annum, respectively in co-operative, public and private sector sugar factories, respectively. Whereas the rates of growth of fixed assets where 7.77, 14.20 and 15.79 per cent in that order, resulting in increased total assets position of co-operatives to Rs. 11491.38 lakhs and private and public sector sugar factories to Rs. 21,608.16 lakhs and Rs. 4042.89 lakhs, respectively during 2004-05.

The current, fixed and total assets were higher in the case of above average factories, as compared to their counterparts in respective sector.

Doodhganga sugar factory (Co-operative) had the highest fixed and total assets, whereas Ugarkhurd sugar factory (Private) had the highest current assets (Rs. 5403.10 lakhs) in 1990-01 to Rs. 16588.43 lakhs in 2004-05) during the study period.

5.2.2.5 *Liabilities position*

The current liabilities (Appendix-16 and 16a), borrowings (Appendix-17 and 17a) and total liabilities (Appendix-18 and 18a) grew from Rs.1, 175.13, Rs. 1340.45 and Rs. 1476.73 lakhs during 1990-91 to Rs.4743.60, Rs.9340.67 and Rs 14084.27 lakhs during 2004-05 respectively with compound growth rates of 8.14, 13.23 and 11. 42 per cent in that order.

Co-operative sector sugar factories had the highest borrowings during the study period followed by Private and public sector sugar factories. The current liabilities increased at a faster rate (12.61 per cent per annum) in the case of public sector sugar factories, followed by private and co-operative (6.08 per cent and 3.55 per cent, respectively) sector sugar factories. This resulted in a 12.18 per cent growth rate of total liabilities per annum over the study period in the public sugar factories, followed by private sugar factories (11.80 per cent). The lowest growth of total liabilities was found in the case of co-operative sector sugar factories (9.77 per cent).

Among the below average sugar factories, the total borrowings were the highest in the case of private sector followed by public and co-operative sugar factories. Where as the current liabilities increased steadily in public sector (growth rate of 11.08 per cent) and private sector (Davanagere, 10.95 per cent) and co-operative sector (8.32 per cent). The total liabilities were the highest with the highest growth rate (17.57 per cent, Davanagere) in Private sector sugar factories, followed by co-operative sector (11.16 per cent) and public sector (7.40 per cent, Mysore Paper Mills Bhadravathi) sugar factories.

Among the above average sugar factories, the total borrowings were the lowest in the co-operative sector sugar factory (Malaprabha Sugar Factory). The annual growth rates of borrowings were the highest in My Sugar Public Sector sugar factory (17.11 per cent) followed by co-operative sector (12.88 per cent) and private sector (12.32 per cent, Ugar sugar factory). Similar trend was observed with respect to current liabilities and total liabilities.

5.2.2.6 *Total sales*

Regarding the sales transaction of sugar factories, the over all sales increased from Rs.3100.77 lakhs in 1990-91 at an annual growth rate of 6.51 per cent to Rs.7677.99 lakhs in 2004-05 (Appendix-19 and 19a).

Total sales were the highest in the case of private sector sugar factories. They increased from Rs.4031.21 lakhs in 1990-91 to Rs.11891.12 lakhs in 2004-05 with an annual increase of sales at the rate of 7.13 per cent followed by co-operative and public sector sugar factories. Similar results were observed in the case of below average sugar factories as well as above average sugar factories.

5.2.2.7 *Gross income and total expenses*

Both gross income (Appendix-.20 and 20a) and total expenses (Appendix-21 and 21a), increased at the rate of 6.97 and 6.44 per cent per annum, respectively. over the study period.

The gross income and the total expenses were the highest in private sector sugar factories over the study period, followed by public and co-operative sector sugar factories.

Similar trends were observed with respect to both gross income and total expenses among the below average sugar factories, whereas among the above average sugar factories, the private sector factory ranked first, followed by public sector and co-operative sugar factories.

5.2.2.8 *Net profit*

Net profits (+)/losses (-) of the sugar factories over the study period are presented in Appendix-22 and 22a.

The net profits were found to be widely fluctuating, the overall average net loss during the study period being Rs. 2.31 lakhs. In the sector-wise analyses, the private sugar factories had an average net profit of Rs. 134.36 lakhs. Both incase of co-operative sugar factories and public sector sugar factories incurred a average loss of Rs. 49.28 and Rs. 92 lakhs respectively over a period of study.

Among the below and above average sugar factories, net profits were the highest in private sector sugar factories (Rs. 124.48 lakhs, Gataprabha) followed by private sector sugar factories (Rs. 329.45 lakhs, Ugar sugars), and public sector sugar factories (Rs.51.14 lakhs, Mysore Paper Mills). The net profit of Rs. 382.47 lakhs was observed in the case of above average co-operative sugar factories where as below average co-operative sugar factories incurred an average loss of Rs. 235.97 lakhs during the period.

5.3 Financial Analyses of Sugar Factories

The following are the financial tools used to analyze the data of the selected sugar factories

5.3.A. Ratio Analyses

5.3.B. Dupont Analyses

5.3.C. Break Even Analyses

5.3.D. Safety Margin

5.3. A. Ratio Analyses

Ratios are measurements of results originated from a business. The ratio analyses is one of the most useful and common methods of analysing the financial statements. A more rigorous financial analyses of the selected sugar factories (individually and collectively) was attempted through selected financial ratios.

Ratio analyses was carried out for the period from 1990-91 to 2004-05. In this chapter, only the weighted average ratios are presented. However, the year wise, factory wise (individually and collectively) ratios are given in Appendices. The selected financial ratios fall under six groups *viz.*, tests of liquidity, tests of solvency, tests of turnover, tests of efficiency and profitability, tests of financial strength and the fixed assets ratios.

5.3.1 *Tests of liquidity*

The liquidity ratios were employed to measure the ability of the sugar factory to meet its immediate maturing obligations out of its own short-term resources. These ratios provide a better insight into liquidity health of an enterprise and help to assess the economic health of an enterprise. Thus, liquidity ratios are treated as inputs to analyze the quality of the capital structure and assets portfolio of an enterprise. In all, four liquidity ratios were worked out and are presented in Table 5.1 and Appendices 23 to 26,

5.3.1.1 *Current ratio*

The ratio of current assets to current liabilities is termed as current ratio. The overall weighted average current ratio for the sugar factories was found to be 2.92 (Appendix-23).

The ratio was found to be the highest in the case of private sector sugar factories (3.56), followed by co-operative sector (3.30) and public sector sugar factories (1.89).

The ratio was greater than unity in all the factories. The ratio was the highest in Doodhganga co-operative sugar factory (4.61, above average co-operative), followed by Ghataprabha sugar factory (3.77, below average co-operative) and Hiranyakeshi co-operative sugar factory (3.40, above average co-operative). Among the above average sugar factories, it was the highest in Doodhganga co-operative sugar factory (4.61) followed by Malaprabha co-operative sugar factory (2.54) and Ugarkhurd sugar factories (2.21, Private).

5.3.1.2 *Acid test ratio*

Acid test ratio, or quick ratio, provides a better measure of liquidity than the current ratio, as inventories are deducted from current assets for calculation of this ratio (Appendix-24).

This ratio was found to be above the optimum ratio level of 1:1 in all the cases. However, it was the highest in the case of Doodhganga sugar factory (4.61, above average co-operative), followed by Ghataprabha sugar factory (3.77, below average co-operative). The lowest ratios were found in My Sugar sugar factory (1.85, above average public) and Mysore Paper Mills, Bhadravathi sugar factory (1.93, below average public).

The overall ratio was found to be satisfactory (2.92) on par with a current ratio of 2.92.

TABLE 5.1

TESTS OF LIQUIDITY FOR SAMPLE SUGAR FACTORIES

SECTORS	*FACTORIES*	*CURRENT RATIO*	*ACID TEST RATIO*	*LIQUID ASSETS TO TOTAL ASSETS*	*INVENTORY TO NET WORKING CAPITAL RATIOS*
ABOVE AVERAGE	HIRANYAKESHI	3.40	3.40	0.64	1.42
CO-OPERATIVE S.F	MALAPRABHA	2.54	2.54	1.10	1.19
	DOODGHANGA	4.61	4.61	2.20	2.40
	AVG	3.52	3.52	1.31	1.67
BELOW AVERAGE	PANDAVAPURA	3.37	3.37	3.28	3.02
CO-OPERATIVE S.F	GHATAPRABHA	3.77	3.77	1.43	1.58
	HEMAVATHI	2.11	2.11	1.68	1.70
	AVG	3.08	3.08	2.13	2.10
OVER ALL CO-OPERATIVE S.F	AVG	3.30	3.30	1.72	1.89
ABOVE AVERAGE	MY SUGAR	1.85	1.85	1.90	2.07
PUBLIC SECTOR S.F	AVG	1.85	1.85	1.85	1.85
BELOW AVERAGE	MYSORE	1.93	1.93	1.49	1.60
PUBLIC SECTOR S.F	AVG	1.93	1.93	1.49	1.60

TABLE 5.1 *contd...*

SECTORS	*FACTORIES*	*CURRENT RATIO*	*ACID TEST RATIO*	*LIQUID ASSETS TO TOTAL ASSETS*	*INVENTORY TO NET WORKING CAPITAL RATIOS*
OVER ALL PUBLIC SECTOR S.F	AVG	1.89	1.89	1.67	2.84
ABOVE AVERAGE	UGAR	2.21	2.21	1.12	1.27
PRIVATE SECTOR S.F	AVG	2.21	2.21	1.12	1.27
BELOW AVERAGE	DHAVANAGERE	2.72	2.72	1.54	1.63
PRIVATE SECTOR S.F	AVG	2.72	2.72	1.54	1.63
OVER ALL PRIVATE SECTOR S.F	AVG	3.56	3.56	1.89	1.50
OVERALL AVERAGE		2.92	2.92	1.76	2.08

Source: *Compiled and Analysed from the annual reports and general ledgers of above sugar factories from 1990-91 to 2004-05*

5.3.1.3 *Liquid Assets to Total Assets Ratio*

The overall liquid assets to total assets ratio was 1.76, indicating that on an average the sugar factory had 176 per cent of total assets as liquid assets (Appendix-25).

This ratio was the lowest in the case of public (1.67), followed by co-operative (1.72) and private sector (1.89) sugar factories.

Below average co-operative sugar factories had the highest ratio (2.13) followed by the private sector (1.89) and public sector (1.67).

5.3.1.4 *Inventory Ratio*

The overall average ratio was found to be greater than the unity indicating that the net working capital was tied up in inventories. However, it was the lowest in private (1.50) sector sugar factory, followed by co-operative sector (1.89) and public sector (2.84) sugar factories (Appendix-26).

5.3.2 ***Tests of solvency***

Three different ratios were developed and were made use of to measure the share of the members of the sugar factories against the funds provided by its creditors. These ratios, presented in Table-5.2 and Appendices 27, 28 and 29 were computed to solicit information on the ability of the sugar factories to cover its medium term and long-term obligations, the margin of safety offered to the creditors and the potential earnings from the use of borrowed funds.

5.3.2.1 *Total Liabilities to Owned Funds Ratio*

The overall average ratio was found to be 7.41, whereas the ratio was beyond the prescribed norms of 3:1 in all sectors of sugar factories (Appendix-27).

Among the below average sugar factories, public sector had the lowest ratio (1.82), followed by co-operative sector (4.69) and private sector sugar factories (7.13).

Among the above average sugar factories, the lowest ratio were found in public sector (6.35) closely followed by private sector (7.69) sugar factories. However, the ratio was as high as (17.58) in the case of co-operative sector (29.66, Hiranyakeshi) sugar factory.

5.3.2.2 *Fixed Assets to Owned Funds Ratio*

Fixed assets to owned funds ratio was lower in below average public sector sugar factories than their counter parts in each group (ownership), the overall average ratio being 2.23 (Appendix-28).

Private sector sugar factories had an average ratio of 2.23, which was lower than the other two sector sugar factories. The ratio was the highest in Hiranyakeshi sugar factory (17.51, above average co-operative), followed by Doodhganga sugar factory (8.02, above average co-operative), My Sugar sugar factory (6.38, above average Public), Pandavapura sugar factory (6.36, below average co-operative) and Ugarkhurd sugar factory (2.74, above average Private).

5.3.2.3 *Debt Equity Ratio*

Debt equity ratio was found to be less in public sector sugar factories (2.23), followed by private sector sugar factories (4.6) and Co-operative sector sugar factories (6.92). It is observed that debt equity ratio was found highest in the Hiranyakeshi sugar factory (18.14, above average co-operative) followed by Pandavapura sugar factory (10.51, below average co-operative). However the overall weighted average was 4.62 (Appendix-29).

5.3.3 ***Tests of turnover***

The ratios, which are known as activity ratios, indicate the efficiency of an organisation in the utilisation of available resources. The turn over ratios *viz.*, inventory turnover ratio, total assets turnover ratio and working capital turnover ratio are presented in Table-5.3 and Appendices 30, 31 and 32.

TABLE 5.2

TESTS OF SOLVENCY FOR SAMPLE SUGAR FACTORIES

SECTORS	*FACTORIES*	*TOTAL LIABILITIES TO OWNED FUNDS*	*FIXED ASSETS TO OWNED FUNDS*	*DEBT EQUITY RATIO*
ABOVE AVERAGE CO-OPERATIVE S.F	HIRANYAKESHI	29.66	17.51	18.14
	MALAPRABHA	17.57	6	7.43
	DOODGHANGA	5.52	8.02	4.42
	AVG	17.58	10.51	10.00
BELOW AVERAGE CO-OPERATIVE S.F	PANDAVAPURA	13.98	6.36	10.51
	GHATAPRABHA	1.62	0.95	1.15
	HEMAVATHI	1.07	1.94	0.54
	AVG	5.56	3.08	4.07
OVER ALL CO-OPERATIVE S.F	AVG	11.57	6.80	7.03
ABOVE AVERAGE PUBLIC SECTOR S.F	MY SUGAR	6.35	6.38	3.4
	AVG	6.35	6.38	3.4
BELOW AVERAGE PUBLIC SECTOR S.F	MYSORE	1.82	1.69	1.06
	AVG	1.82	1.69	1.06
OVER ALL PUBLIC SECTOR S.F	AVG	4.09	4.04	2.23

TABLE 5.2 *contd...*

SECTORS	FACTORIES	TOTAL LIABILITIES TO OWNED FUNDS	FIXED ASSETS TO OWNED FUNDS	DEBT EQUITY RATIO
ABOVE AVERAGE	UGAR	7.69	2.74	4.52
PRIVATE SECTOR S.F	AVG	7.69	2.74	4.52
BELOW AVERAGE	DHAVANAGERE	7.13	1.72	4.69
PRIVATE SECTOR S.F	AVG	7.13	1.72	4.69
OVER ALL PRIVATE SECTOR S.F	AVG	7.41	2.23	4.605
OVER ALL AVERAGE		7.69	4.35	4.62

Source: *Compiled and Analysed from the annual reports and general ledgers of above sugar factories from 1990-91 to 2004-05*

5.3.3.1 *Inventory Turn over Ratio*

The inventory turnover ratio was higher in above average sugar factories than below average factories in all the sectors. The average inventory turnover ratio was found to be 2.80 (Appendix-30).

Among the above average sugar factories, co-operative sector sugar factories had the highest inventory turnover ratio (8.92) followed by public (2.54) and private sector sugar factories (2.05), whereas among the below average sugar factories, private sector sugar factories had the highest inventory turnover ratio (1.14) followed by public sector (1.12) and co-operatives (1.05).

5.3.3.2 *Total Assets Turnover Ratio*

This ratio was greater than one in public sector (2.02) and co-operative sector Sugar factories (1.04), whereas in private sector it was less than one (0.9). This ratio was highest in the My Sugar factory (3.41, above average public sector) (Appendix-31).

5.3.3.3 *Working Capital Turnover Ratio*

Working capital turnover ratio was the highest in public sector (6.14) followed by co-operative sector (3.8) and private sector (2.87) (Appendix-32).

Among the below average sugar factories, public sector registered the highest ratio of 8.13, followed by private sector (2.49) and co-operative sector (1.57).

Among the above average sugar factories, co-operative sector sugar factories had the highest ratio (6.04), followed by public sector (4.15) sugar factories and private sector (3.25). The ratio was greater than one in all the sugar factories the overall weighted average being 4.28.

TABLE 5.3

TESTS OF TURNOVER FOR SAMPLE SUGAR FACTORIES

SECTORS	FACTORIES	INVENTORY TURNOVER RATIO	TOTAL ASSETS TURNOVER RATIO	WORKING CAPITAL TURNOVER RATIO
ABOVE AVERAGE CO-OPERATIVE S.F	HIRANYAKESHI	1.37	0.67	1.88
	MALAPRABHA	23.4	0.73	2.49
	DOODGHANGA	1.98	0.54	13.74
	AVG	8.92	0.65	6.04
BELOW AVERAGE CO-OPERATIVE S.F	PANDAVAPURA	0.92	0.23	1.09
	GHATAPRABHA	1.53	0.76	1.84
	HEMAVATHI	0.71	3.29	1.91
	AVG	1.05	1.43	1.61
OVER ALL CO-OPERATIVE S.F	AVG	4.99	1.04	3.83
ABOVE AVERAGE PUBLIC SECTOR S.F	MY SUGAR	2.54	3.41	4.15
	AVG	2.54	3.41	4.15
BELOW AVERAGE PUBLIC SECTOR S.F	MYSORE	1.12	0.62	8.13
	AVG	1.12	0.62	8.13
OVER ALL PUBLIC SECTOR S.F	AVG	1.83	2.015	6.14

TABLE 5.3 *contd...*

SECTORS	*FACTORIES*	*INVENTORY TURNOVER RATIO*	*TOTAL ASSETS TURNOVER RATIO*	*WORKING CAPITAL TURNOVER RATIO*
ABOVE AVERAGE	UGAR	2.05	1	3.25
PRIVATE SECTOR S.F	AVG	2.05	1	3.25
BELOW AVERAGE	DHAVANAGERE	1.14	0.8	2.49
PRIVATE SECTOR S.F	AVG	1.14	0.8	2.49
OVER ALL PRIVATE SECTOR S.F	AVG	1.59	0.9	2.87
OVER ALL AVERAGE		2.80	1.32	4.28

Source: *Compiled and Analysed from the annual reports and general ledgers of above sugar factories from 1990-91 to 2004-05.*

5.3.4 *Tests of efficiency and profitability*

The profitability level and degree of efficient utilisation of the resources were measured by employing different ratios. The efficiency ratios *viz.*, efficiency of capital ratio and gross ratio and the profitability ratios *viz.*, net profits to owned funds, net profits to total assets and net profits to total sales ratios are presented in Table 5.4 and Appendix-33, 34, 35, 36 and 37.

5.3.4.1 *Efficiency of Capital Ratio*

An overall weighted average efficiency of capital ratio 0.25 was observed. The public sector sugar factories registered the highest ratio (0.39) followed by the co-operatives (0.19) and private sector (0.17) (Appendix-33).

Among above average sugar factories public sector has the highest ratio (0.64) followed by the co-operative sector (0.1) and private sector (0.03). Among below average sugar factory next to public sector, co-operative sector has the highest ratio (0.43) followed by the private sector (0.32).

5.3.4.2 *Gross Ratio (Percentage)*

The overall weighted average ratio was 292.87 per cent indicating that the factory on an average incurred total expenses of Rs. 292.87 for every Rs. 100.00 gross income (Appendix-34).

The ratio was more than 100 per cent in co-operative sector (651.03) followed by public sector (131.4) this ratio was highest in Pandavapura sugar factory (3118.77, below average co-operative).

The ratio was the lowest, in the case of below average private Sugar factory (94.38 per cent) followed by above average private sugar factory (97.97 per cent) and among the factories under different ownerships, the lowest ratio was Found in the case of private sector (96.18) followed by public (131.4 per cent) and co-operatives (651.03 per cent).

TABLE 5.4

TESTS OF EFFICIENCY AND PROFITABILITY FOR SAMPLE SUGAR FACTORIES

SECTORS	*FACTORIES*	*EFFICIENCY OF CAPITAL RATIO*	*GROSS RATIO (%)*	*NET PROFITS TO OWNED FUNDS RATIO*	*NET PROFITS TO TOTAL ASSETS RATIO*	*NET PROFITS TO TOTAL SAELS RATIO*
ABOVE AVERAGE CO-OPERATIVE S.F	HIRANYAKESHI	0.03	123.85	25.1	1	1.35
	MALAPRABHA	0.07	193.25	-16.18	-1.36	-2.59
	DOODGHANGA	0.19	49.57	30.79	2.53	5.61
	AVG	0.10	122.22	13.24	0.72	1.46
BELOW AVERAGE CO-OPERATIVE S.F	PANDAVAPURA	0.43	3118.77	-190.41	-7.97	-54.87
	GHATAPRABHA	0.17	169.53	6.25	3.06	3.06
	HEMAVATHI	0.27	251.19	1.56	16.11	1.04
	AVG	0.29	1179.83	-60.87	3.73	-16.92
OVER ALL CO-OPERATIVE S.F	AVG	0.19	651.03	-23.82	2.23	-7.73
ABOVE AVERAGE PUBLIC SECTOR S.F	MY SUGAR	0.14	104.84	-14.36	-6.69	-1.38
	AVG	0.14	104.84	-14.36	-6.69	-1.38
BELOW AVERAGE PUBLIC SECTOR S.F	MYSORE	0.64	157.96	1.67	1.12	3.02
	AVG	0.64	157.96	1.67	1.12	3.02
OVER ALL PUBLIC SECTOR S.F	AVG	0.39	131.4	-6.345	-2.785	0.82

TABLE 5.4 *contd...*

SECTORS	FACTORIES	EFFICIENCY OF CAPITAL RATIO	GROSS RATIO (%)	NET PROFITS TO OWNED FUNDS RATIO	NET PROFITS TO TOTAL ASSETS RATIO	NET PROFITS TO TOTAL SAELS RATIO
ABOVE AVERAGE	UGAR	0.03	97.97	16.19	2.01	1.97
PRIVATE SECTOR S.F	AVG	0.03	97.97	16.19	2.01	-0.21
BELOW AVERAGE	DHAVANAGERE	0.32	94.38	5.49	0.99	-0.21
PRIVATE SECTOR S.F	AVG	0.32	94.38	5.49	0.99	
OVER ALL PRIVATE SECTOR S.F AVG		0.175	96.175	10.84	1.5	0.88
OVER ALL AVERAGE		0.25	292.87	-6.44	0.31	-2.01

5.3.4.3 *Net Profits to Owned Funds Ratio*

The net profits to owned funds ratio was found to be negative in the case of co-operative sugar factories (-23.82) followed by public sector (-6.34). Whereas this ratio was the highest in the case of private sugar factories (10.84), indicating a higher rate of return on owned funds (Appendix-35).

5.3.4.4 *Net Profits to Total Assets Ratio*

This ratio was negative in case of Mysugar (-6.69, above average public sector), followed by Pandavapura sugar factory (-7.97, below average co-operative) and Malaprabha (-1.36, above average co-operative). The ratio was the highest in the case of co-operative sugar factory (2.23), followed by private sector sugar factory (1.5) (Appendix-36).

The overall weighted average ratio was 0.31, indicating net profit earnings of Rs. 31 for every Rs. 100.00 total assets. The ratio was the highest in the case of below average co-operative sugar factory (3.73), followed by above average private sugar factory (2.01).

5.3.4.5 *Net Profits to Total Sales Ratio*

The net profit to total sales ratio was negative in Pandavapura sugar factory (-54.87, below average co-operative), followed by Malaprabha sugar factory (-2.59, above average co-operative) and Mysugar (-1.38, above average public) (Appendix-37).

The ratio was greater in Doodhganga sugar factory (5.61, above average co-operative), followed by Mysore sugar factory (3.02, below average public) and Ugar sugar factory (1.97, above average private sector).

TABLE 5.5

TESTS OF FINANCIAL STRENGTH FOR SAMPLE SUGAR FACTORIES

SECTORS	*FACTORIES*	*NET WORTH (RS IN LAKH)*	*NET CAPITAL RATIO*	*EQUITY CAPITAL RATIO*
ABOVE AVERAGE CO-OPERATIVE S.F	HIRANYAKESHI	6957.95	1.72	0.98
	MALAPRABHA	3172.72	1.56	1
	DOODGHANGA	7382.82	2.48	0.59
	AVG	5837.83	1.92	0.86
BELOW AVERAGE CO-OPERATIVE S.F	PANDAVAPURA	3131.12	1.8	1
	GHATAPRABHA	1920.39	1.57	1
	HEMAVATHI	1346.46	0.71	1
	AVG	2132.66	1.36	1.00
OVER ALL CO-OPERATIVE S.F	AVG	3985.24	1.64	0.93
ABOVE AVERAGE PUBLIC SECTOR S.F	MY SUGAR	-4315.44	0.64	1
	AVG	-4315.44	0.64	1
BELOW AVERAGE PUBLIC SECTOR S.F	MYSORE	41378.26	14.46	1
	AVG	41378.26	14.46	1
OVER ALL PUBLIC SECTOR S.F AVG		18531.41	7.55	1

TABLE 5.5 *contd...*

SECTORS	FACTORIES	NET WORTH (RS IN LAKH)	NET CAPITAL RATIO	EQUITY CAPITAL
ABOVE AVERAGE	UGAR	2200.7	1.17	1
PRIVATE SECTOR S.F	AVG	2200.7	1.17	1
BELOW AVERAGE	DHAVANAGERE	519.8	1.06	1
PRIVATE SECTOR S.F	AVG	519.8	1.06	1
OVER ALL PRIVATE SECTOR S.F AVG		1360.25	1.115	1
OVER ALL AVERAGE		7958.97	3.44	0.98

Source: *Compiled and Analysed from the annual reports and general ledgers of above sugar factories from 1990-91 to 2004-05.*

TABLE 5.6

TESTS OF FIXED ASSETS RATIOS FOR SAMPLE SUGAR FACTORIES

SECTORS	*FACTORIES*	*FIXED ASSETS TO TOTAL ASSETS*	*FIXED ASSETS TO NET WORTH*	*FIXED ASSETS TO TOTAL SALES*
ABOVE AVERAGE CO-OPERATIVE S.F	HIRANYAKESHI	0.36	0.61	0.55
	MALAPRABHA	0.25	0.88	0.38
	DOODGHANGA	0.72	1.32	1.41
	AVG	0.44	0.94	0.78
BELOW AVERAGE CO-OPERATIVE S.F	PANDAVAPURA	0.37	1.07	2.24
	GHATAPRABHA	0.38	1.12	0.51
	HEMAVATHI	4.7	2.34	1.48
	AVG	1.82	1.51	1.41
OVER AL I L CO-OPERATIVE S.F	AVG	1.13	1.22	1.10
ABOVE AVERAGE PUBLIC SECTOR S.F	MY SUGAR	5.67	-2.79	0.8
	AVG	5.67	-2.79	0.8
BELOW AVERAGE PUBLIC SECTOR S.F	MYSORE	0.72	-2.07	1.55
	AVG	0.72	-2.07	1.55
OVER ALL PUBLIC SECTOR S.F	AVG	3.195	-2.43	1.175

TABLE 5.6 *contd...*

SECTORS	*FACTORIES*	*FIXED ASSETS TO TOTAL ASSETS*	*FIXED ASSETS TO NET WORTH*	*FIXED ASSETS TO TOTAL SALES*
ABOVE AVERAGE	UGAR	0.33	1.71	0.36
PRIVATE SECTOR S.F	AVG	0.33	1.71	0.36
BELOW AVERAGE	DHAVANAGERE	0.22	2.25	0.55
PRIVATE SECTOR S.F	AVG	0.22	2.25	0.55
OVER ALL PRIVATE SECTOR S.F	AVG	0.28	1.98	0.46
OVER ALL AVERAGE		1.53	0.26	0.91

Source: *Compiled and Analysed from the annual reports and general ledgers of above sugar factories from 1990-91 to 2004-05*

TABLE. 5.7

DUPONT ANALYSES FOR SELECTED SUGAR FACTORIES IN KARNATAKA FROM 1990-91 TO 2004-05

YEARS	*CO-OPERATIVE SECTOR*								*Avg*
	Above Average				*Below Average*				
	Hiranya-keshi	*Mala-prabha*	*Dood-ghanga*	*Avg*	*Panda-vapura*	*Ghata-prabha*	*Hema-vathi*	*Avg*	
1990 - 91	0.25	0.01	0.06	0.10	0.00	0.12	0.90	0.34	0.22
1991 - 92	0.15	0.01	0.01	0.06	-0.22	0.39	0.65	0.27	0.17
1992 - 93	0.15	0.02	0.05	0.07	0.00	-1.17	0.26	-0.30	-0.12
1993 - 94	0.16	0.01	0.02	0.06	0.00	0.73	0.02	0.25	0.16
1994 - 95	0.09	0.01	0.99	0.36	-0.26	1.02	0.06	0.27	0.32
1995 - 96	1.33	0.00	1.13	0.82	-0.09	0.03	0.00	-0.02	0.40
1996 - 97	0.06	0.00	0.95	0.34	-0.45	0.23	-0.29	-0.17	0.08
1997 - 98	0.08	0.00	0.39	0.16	-0.54	0.44	-0.03	-0.04	0.06
1998 - 99	0.10	0.00	1.68	0.59	-6.55	-0.34	-0.52	-2.47	-0.94
1999 - 00	0.10	0.00	0.56	0.22	-4.46	0.50	-0.11	-1.36	-0.57
2000 - 01	0.11	0.01	0.99	0.37	-3.95	0.90	-0.15	-1.06	-0.35
2001 - 02	0.09	0.00	0.18	0.09	-4.96	0.08	-0.06	-1.65	-0.78
2002 - 03	0.10	-0.85	0.72	-0.01	-4.43	-0.13	-0.50	-1.69	-0.85
2003 - 04	1.11	-0.89	0.33	0.18	-2.99	-0.25	0.02	-1.07	-0.44
2004 05	1.98	-0.83	0.26	0.47	0.00	0.59	0.39	0.33	0.40

Source: *Compiled and Analysed from the annual reports and general ledgers of above sugar factories from 1990-91 to 2004-05*

TABLE 5.7a

DUPONT ANALYSES FOR SELECTED SUGAR FACTORIES IN KARNATAKA FROM 1990-91 TO 2004-05

YEARS	PUBLI/JOINT SCTOR			PRIVATE SECTOR			Overall
	Above Avg My Sugar	*Below Avg Mysore*	*Avg*	*Above Avg Ugar*	*Below Avg Dhavanagere*	*Avg*	*Avg*
1990 - 91	0.54	0.00	0.27	0.39	0.17	0.28	0.25
1991 - 92	-0.10	0.20	0.05	0.45	0.13	0.29	0.17
1992 - 93	-0.82	0.00	-0.41	0.47	0.24	0.35	-0.07
1993 - 94	0.81	0.06	0.44	0.34	0.31	0.33	0.27
1994 - 95	1.31	0.25	0.78	0.91	0.31	0.61	0.51
1995 - 96	0.28	0.33	0.30	1.44	0.05	0.75	0.46
1996 - 97	-0.03	0.02	-0.01	0.88	0.02	0.45	0.15
1997 - 98	-0.36	0.00	-0.18	0.81	0.12	0.47	0.10
1998 - 99	0.21	0.01	0.11	0.80	0.06	0.43	-0.33
1999 - 00	0.04	0.01	0.03	1.80	0.09	0.94	-0.04
2000 - 01	-0.23	0.07	-0.08	2.98	0.11	1.54	0.19
2001 - 02	-0.10	-0.29	-0.20	0.09	0.01	0.05	-0.43
2002 - 03	-2.96	-0.18	-1.57	-0.23	0.01	-0.11	-0.84
2003 - 04	-143	-0.08	-0.76	0.13	-0.22	-0.04	-0.42
2004 - 05	-0.02	0.05	0.01	0.10	-0.39	-0.15	0.17

Source: *Compiled and Analysed from the annual reports and general ledgers of above sugar factories from 1990-91 to 2004-05*

5.3.B. Dupont Analyses

The Dupont Formula allows examining how a firm generates its return on equity. It links several critical ratios, If operating management is strong (high PM), and asset management is strong (high TAT), and capital structure management is strong (appropriately low EM), then ROE will be high for sound business reasons and the firm can be said to be in strong financial condition.

Return on Equity represents the profitability of funds invested by the owners of the firm. All firms should attempt to make ROE as high as possible over the long-term. ROE can be high for the wrong reasons however. For example, when ROE is high because the equity multiplier is high, this means that high returns are really coming from overuse of debt, which can spell trouble for the firm. Most of ROE should be produced by high ROI (Return on Investment), PM (Profit Margin), and TAT (Total Asset Turnover) and not from high EM (Equity Multiplier)

The results of the Dupont analyses are presented in Tables 5.7 and 5.7a and Graph 1, 2 and 3.

- **Co-operative Sector**

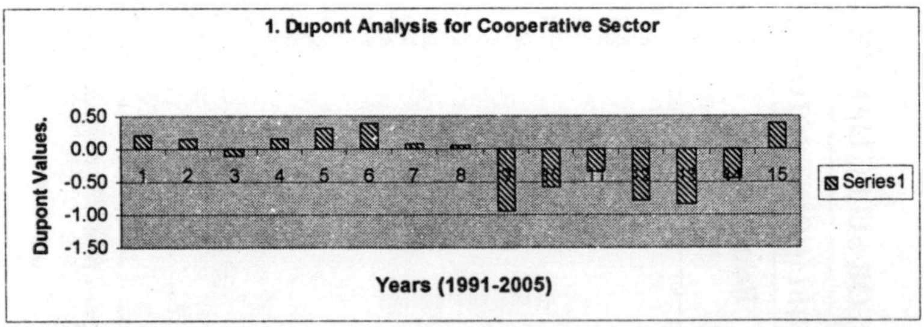

Return on equity of co-operative sugar factories had been negative till the year 2003-04 from the year 1998-99 this is due to weakness of operating management and asset management and the capital structure of co-operativé sector is weak. Here the equity multiplier is high and there is over use of debt during 1991-1998 but in the year 2004-05 return on equity is showing positive trend because of improvement in asset management and capital structure of the sector.

- **Public Sector**

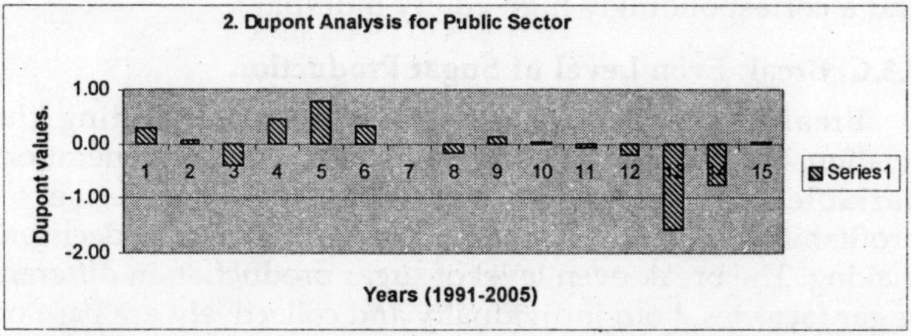

Return on equity of public sector sugar factories has been negative till the year 2003-04 from the year 2000-01. Public sector did well in the beginning that is in the year 1991-92 and after that earned negative return on equity for 2 years and increased its return on equity for continuous three years and ROE falls continuously till the year 2003-04 this is due to weakness of operating management and asset management and the capital structure. But in the year 2004-05 return on Equity is showing positive trend because of improvement in asset management and capital structure of the sector.

- **Co-operative Sector**

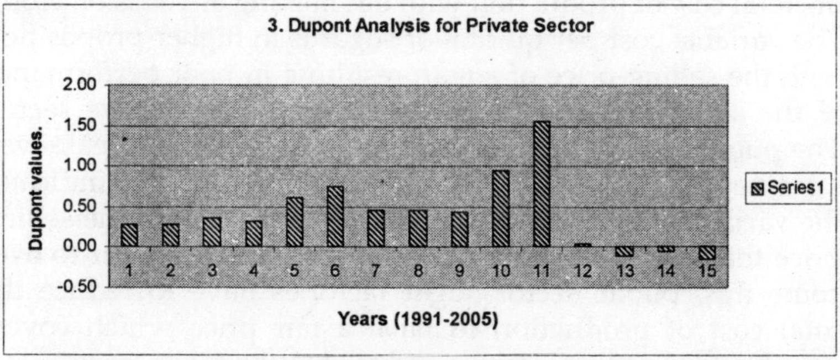

Return on Equity of Private sector factories had been Positive till the year 2001-02 from the year 1991-92 this is due to strong in operating management and asset management and the capital structure. But from the year 2002-03 onwards Return on Equity is showing negative trend because of weak asset management and capital structure and high levels of debt. Debt requires fixed payments of principle and interest and a correspondingly high equity multiplier.

5.3.C. Break-Even Level of Sugar Production

Break-Even Analysis is useful in understanding the relationship between level of operation, revenue generation, variable cost and fixed costs and their influence on profitability. As such, it has a wide variety of uses in decision-making. The break-even level of sugar production in different sugar factories, both individually and collectively are base on ownership. The results of the breakeven analysis are presented in the tables 5.8 (calculated for the year 2004-05), 5.9, 5.9a and Graphs 4 to 7. To be at the break-even point, on an average, with the existing cost structure it is found that a sugar factory of an above average co-operative sector as to produce an average of 1,68,804 quintals of sugar to attain BEP (Break Even Point) where in case of the below average co-operative sugar factories an average has to be 1,09,882 quintals of sugar production to achieve BEP. Comparatively below average co-operative sugar factories posses lesser break-even sales than that of above average co-operative sugar factories which indicates the level of sugar production just sufficient to meet the total cost of production with out making any loss or profit. The variable cost per quintal of sugar is in higher proposition with the selling price of sugar, resulting in poor performance of the sugar factories of below average co-operative sector. The public sector sugar factories have determined break-even sales negatively at –5,61,250 quintal's of sugar; this indicates the variable cost per quintal of sugar is more than the selling price that leads to negative break even sales. In order to over come this, public sector sugar factories have to reduce the total cost of production to have a fair price which covers variable cost and thus loss is reduced to the nil i.e. zero.

The private sector sugar factories on an average have to produce 20,06,904 quintals of sugar to achieve break-even sales. Private sector of sugar factories were on safer side in the beginning by covering the variable cost. And reducing the total cost of production but now the variable cost is more than the selling price led to negative BEP sales.

TABLE 5.8

BREAK EVEN ANALYSES FOR SAMPLE SUGAR FACTORIES DURING 2004-05

SECTORS	*FACTORIES*	*Total Fixed Cost (Rs. in Lakh)*	*Total Variable Cost (Rs. in Lakh)*	*Variable Cost Per Quintal of Sugar (Rs.)*	*Price of Sugar (Rs./Qtl)*	*Break-Even Level of Sugar Production (Rs./Qtls)*
ABOVE AVERAGE CO-OPERATIVE S.F	HIRANYAKESHI	346.59	7148.05	1381.40	1491.83	313853.13
	MALAPRABHA	869.74	3259.28	1437.06	1148.34	-301240.06
	DOODGHANGA	1898.41	1597.64	1360.64	1745.09	493798.93
	AVG	1038.25	4001.66	1393.03	1461.75	168804.00
BELOW AVERAGE CO-OPERATIVE S.F	PANDAVAPURA	2047.94	1078.87	16.80	1142.14	181983.96
	GHATAPRABHA	180.35	5378.35	1114.13	1450.00	53696.87
	HEMAVATHI	68.40	758.30	1221.10	1293.89	93966.04
	AVG	765.56	2405.18	784.01	1295.34	109882.29
OVER ALL CO-OPERATIVE S.F AVG		1331.99	2001.41	1072.33	1520.22	139343.14
ABOVE AVERAGEPUBLIC SECTOR S.F	MY SUGAR	7663.46	7474.66	2083.18	1685.09	-1925058.15
	AVG	7663.46	7474.66	2083.18	1685.09	-1925058.15
BELOW AVERAGEPUBLIC SECTOR S.F	MYSORE	1402.47	6148.14	1505.31	1680.06	802556.19
	AVG	1402.47	6148.14	1505.31	1680.06	802556.19

TABLE 5.8 *contd...*

SECTORS	*FACTORIES*	*Total Fixed Cost (Rs. in Lakh)*	*Total Variable Cost (Rs. in Lakh)*	*Variable Cost Per Quintal of Sugar (Rs.)*	*Price of Sugar (Rs./Qtl)*	*Break-Even Level of Sugar Production (Rs./Qtls)*
OVER ALL PUBLIC SECTOR S.F AVG		4532.96	6811.39	1794.245	1682.57	-561250.98
ABOVE AVERAGEPRIVATE	UGAR	6704.57	28353.41	1496.59	1215.83	-2388006.26
SECTOR S.F	AVG	6704.57	28353.41	1496.59	1215.83	-2388006.26
BELOW AVERAGEPRIVATE	DHAVANAGERE	395.56	1741.09	1282.76	1258.43	-1625803.28
SECTOR S.F	AVG	395.56	1741.09	1282.76	1258.43	-1625803.28
OVEER ALL PRIVATE SECTOR S.F AVG		3550.06	15047.25	1389.68	1237.13	-2006904.77
OVER ALL AVERAGE		3138.34	7953.35	1418.75	1479.97	-809604.20

Source: *Compiled and Analysed from the annual reports and general ledgers of above sugar factories from 1990-91 to 2004-05*

TABLE 5.9

BREAK EVEN ANALYSES FOR SUGAR PRODUCTION FROM 1990-91 TO 2004-05 (QUINTALS)

YEARS	*CO-OPERATIVE SECTOR*								
	Above Average				*Below Average*				*AVG*
	HIRANYA-KESHI	*MALA-PRABHA*	*DOOD-GHANGA*	*AVG*	*PANDA-VAPURA*	*GHATA-PRABHA*	*HEMA-VATHI*	*AVG*	
1990 - 91	-797043.38	549152.78	444152.52	65420.64	117814.55	80375.34	-205507.65	-2439.25	31490.69
1991 - 92	-9412523.94	1552926.20	420369.09	-2479742.88	88122.13	62158.19	463305.69	204528.67	-1137607.11
1992 - 93	-1245441.36	-5771414.16	422324.59	-2198176.98	94594.22	88812.51	1513058.60	565488.44	-816344.27
1993 - 94	-203281.02	-247722.61	-93899.71	-181634.44	62712.69	25619.22	-88595.07	-87.72	-90861.08
1994 - 95	707191.79	3742451.89	634590.40	1694744.69	71348.29	-167671.45	-41603.63	-45975.60	824384.55
1995 - 96	-3960432.23	1592435.69	916599.13	-483799.14	104458.23	71160.18	70282.59	81967.00	-200916.07
1996 - 97	-4005195.19	-6124145.45	1343757.27	-2928527.79	64644.36	41757.77	16263.32	40888.48	-1443819.65
1997 - 98	2128215.49	11778135.36	893190.42	4933180.43	0.00	23273.83	8553.11	10608.98	2471894.70
1998 - 99	675209.33	1095850.87	1703436.68	1158165.62	202959.57	133966.83	123319.21	153415.20	655790.41
1999 - 00	1280340.54	3069597.55	626364.95	1658767.68	233722.69	91355.99	62206.37	129095.02	893931.35
2000 - 01	2954750.05	2109259.64	-4486801.74	192402.65	198695.41	235397.21	117848.07	183980.23	188191.44
2001 -02	941040.82	-1204211.76	420642.43	52490.50	204301.89	160482.87	452130.62	272305.12	162397.81
2002 -03	683271.79	142335.92	677553.07	501053.59	181983.96	30494.57	192815.18	135097.90	318075.75
2003 -04	-833155.89	-412590.96	-1779464.78	-1008403.88	0.00	139631.13	460516.84	200049.32	-404177.28
2004 - 05	313853.13	-301240.06	493798.93	168804.40	0.00	53696.87	93966.04	49220.97	109012.48

Source: *Compiled and Analysed from the annual reports and general ledgers of above sugar factories from 1990-91 to 2004-05*

TABLE 5.9a

BREAK EVEN ANALYSES FOR SUGAR PRODUCTION FROM 1990-91 TO 2004-05 (QUINTALS)

YEARS	PUBLIC SECTOR			PRIVATE SECTOR			OVERALL AVG
	ABOVE AVG	BELOW AVG	AVG	ABOVE AVG	BELOW AVG	AVG	
	MY SUGAR	MYSORE		UGAR	DHAVANAGERE		
1990 - 91	-234970590.91	-104834.59	-117537712.75	-2925559.39	365762.37	-1279898.51	-624203.91
1991 - 92	107370996.19	-138026.22	53616484.99	1930400.86	176155.74	1053278.30	-42164.40
1992 - 93	-832506.68	-118371.10	-475438.89	942663.35	281314.72	611989.04	-102177.62
1993 - 94	3576476.71	-119799.23	1728338.74	5366687.54	362021.13	2864354.34	1386746.63
1994 - 95	-2296640.38	-193497.63	-1245069.00	13026646.66	315284.41	6670965.54	3747675.04
1995 - 96	-793572.61	-268578.51	-531075.56	5093345.88	-1550089.46	1771628.21	785356.07
1996 - 97	-321682.47	-418533.16	-370107.82	4165360.42	5405600.78	4785480.60	1670830.47
1997 - 98	-881038.82	-625673.86	-753356.34	1904176.05	-1400380.40	251897.82	1361896.26
1998 - 99	-1315335.11	-489601.84	-902468.48	1317029.39	129342.31	723185.85	689488.13
1999 - 00	-506740.84	-271059.98	-388900.41	2599680.97	-1367860.42	615910.27	754920.81
2000 - 01	-659741.86	-419057.47	-539399.67	1238718.88	4888830.65	3063774.77	1625983.10
2001 - 02	-260058.46	-675512.68	-467785.57	2013454.04	-2467660.88	-227103.42	-32352.81
2002 - 03	-275093.58	-534056.02	-404574.80	-14635400.74	9217312.87	-2709043.94	-1195484.09
2003 - 04	-88707.84	-430139.34	-259423.59	-2470732.37	-16456947.83	-9463840.10	-4934008.69
2004 - 05	-645561.79	-230203.20	-437882.49	-2388006.26	-1625803.28	-2006904.77	-948946.14

Source: *Compiled and Analysed from the annual reports and general ledgers of above sugar factories from 1990-91 to 2004-05*

TABLE 5.10

SAFETY MARGIN OF SUGAR PRODUCTION FROM 1990-91 TO 2004-05 (QUINTALS)

YEARS	*CO-OPERATIVE SECTOR*								
	Above Average				*Below Average*				*AVG*
	HIRANYA-KESHI	*MALA-PRABHA*	*DOOD-GHANGA*	*AVG*	*PANDA-VAPURA*	*GHATA-PRABHA*	*HEMA-VATHI*	*AVG*	
1990 - 91	18906.86	-16729.07	-16300.64	-4707.62	-11411.53	-5798.98	18956.68	582.06	-2062.78
1991 - 92	181843.87	-39222.85	-14431.69	42729.77	-5110.91	-2896.93	-33154.07	-13720.64	14504.57
1992 - 93	22308.46	113601.48	-13119.37	40930.19	0.00	-4227.55	-127926.76	-44051.43	-1560.62
1993 - 94	2308.77	3824.89	2396.94	2843.53	-2979.01	-869.85	4572.33	241.16	1542.34
1994 - 95	-10376.50	-59008.36	-16697.68	-28694.18	-38069.48	6915.50	2377.06	-9592.31	-19143.24
1995 - 96	43710.54	-23442.22	-21338.80	-356.83	-4578.27	-1850.84	-3541.79	-3323.63	-1840.23
1996 - 97	39635.27	79645.72	-23301.37	31993.21	-6632.85	-949.22	-578.77	-2720.28	14636.46
1997 - 98	-16970.20	-142345.68	-14478.25	-57931.38	100.00	-504.36	-226.71	-210.36	-29070.87
1998 - 99	-4677.55	-18039.95	-21193.87	-14637.12	-7798.11	-3716.52	-6565.07	-6026.57	-10331.85
1999 - 00	-9332.84	-38969.06	-7921.13	-18741.01	-8474.65	-2201.94	-2659.49	-4445.36	-11593.19
2000 - 01	-19418.74	-33696.93	50852.28	-754.46	-8279.14	-4550.90	-4724.86	-5851.63	-3303.05
2001 - 02	-7359.57	18841.87	-4540.68	2313.87	-18844.30	-2933.73	-16990.17	-12922.73	-5304.43
2002 - 03	-4396.81	-1666.36	-7906.19	-4656.45	-9930.96	-347.79	-8313.04	-6197.26	-5426.86
2003 - 04	7623.48	8180.43	24932.10	13578.67	100.00	-3941.48	-20867.75	-8236.41	2671.13
2004 - 05	-1993.69	5266.00	-5286.64	-671.45	0.00	-924.41	-3321.58	-1415.33	-1043.39

Source: *Compiled and Analysed from the annual reports and general ledgers of above sugar factories from 1990-91 to 2004-05.*

TABLE 5.10a

SAFETY MARGIN OF SUGAR PRODUCTION FROM 1990-91 TO 2004-05 (QUINTALS)

YEARS	PUBLIC SECTOR			PRIVATE SECTOR			OVERALL AVG
	ABOVE AVG	BELOW AVG	AVG	ABOVE AVG	BELOW AVG	AVG	
	MY SUGAR	MYSORE		UGAR	DHAVANAGERE		
1990 - 91	6847300.63	6462.36	3426881.50	40210.17	-54670.78	-7230.31	-4646.54
1991 - 92	-2929539.54	6184.93	-1461677.31	-26194.01	-13068.02	-19631.01	-2563.22
1992 - 93	24554.92	4475.72	14515.32	-11839.05	-21037.49	-16438.27	-8999.45
1993 - 94	-69523.79	3220.01	-33151.89	-59343.74	-15347.58	-37345.66	-17901.66
1994 - 95	36309.76	7345.77	21827.77	-79999.21	-198798.16	-139398.69	-79270.96
1995 - 96	13806.25	6761.83	10284.04	-34060.44	72276.61	19108.08	8633.93
1996 - 97	5732.74	9019.33	7376.03	-25313.35	-209378.42	-117345.89	-51354.71
1997 - 98	14228.70	10449.36	12339.03	-10925.39	42420.80	15747.70	-6661.58
1998 - 99	19865.00	11856.44	15860.72	-9735.91	-4641.40	-7188.65	-8760.25
1999 - 00	7458.48	5444.38	6451.43	-12729.89	49018.42	18144.27	3275.54
2000 - 01	6360.40	8048.03	7204.21	-6402.40	-91360.38	-48881.39	-26092.22
2001 - 02	2721.31	9945.39	6333.35	-10359.12	62922.06	26281.47	10488.52
2002 - 03	2352.67	6766.61	4559.64	90742.16	-115171.90	-12214.87	-8820.86
2003 - 04	1608.56	11366.92	6487.74	13564.95	352783.42	183174.19	92922.66
2004 - 05	72270.93	6129.86	39200.40	12387.41	57318.97	34853.19	16904.90

Source: *Compiled and Analysed from the annual reports and general ledgers of above sugar factories from 1990-91 to 2004-05*

I. Co-operative Sector

- **Above Average: (Hiranyakeshi, Malaprabha, Doodghanga)**

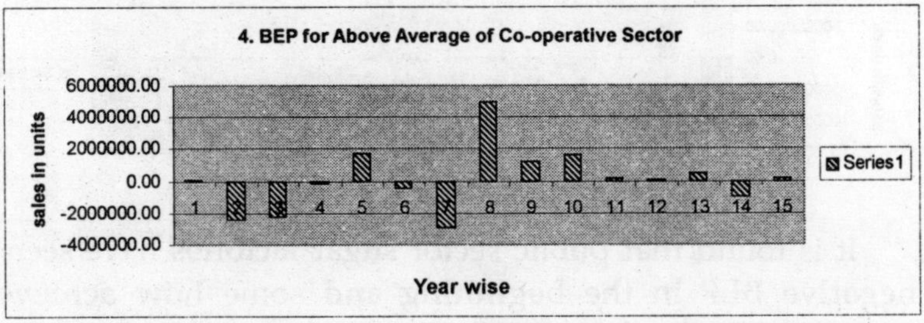

It is observed from the Tables 5.9 and 5.9a that Co-operative sector of sugar factories incurred negative BEP sales because the constant increase in variable cost of production over selling price of sugar. Only in few years it achieved positive BEP sales due to abandant sugarcane availability.

- **Below Average: (Pandavapura, Ghataprabha, Hemavathi)**

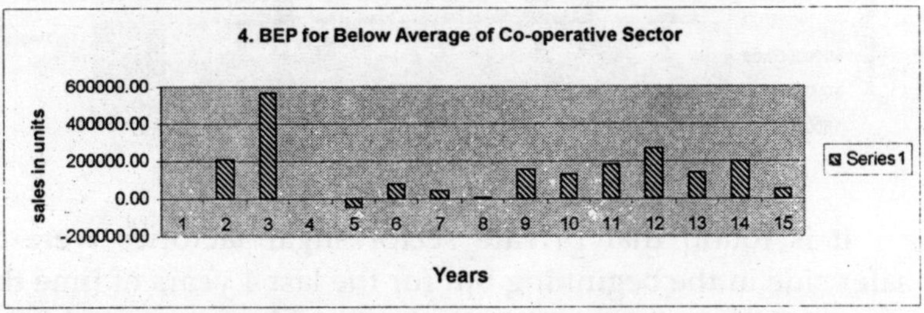

Although below average co-operative sector of sugar factories has achieved BEP sales in which they are all incurring no loss no profit. The above diagram explains that there is fluctuation of income to the above sugar factories over 15 years of time. In the beginning this sector should have produced more volume of sugar to achieve BEP level. But for the last year it is found that BEP sales is less compared to last six years.

II. Public Sector

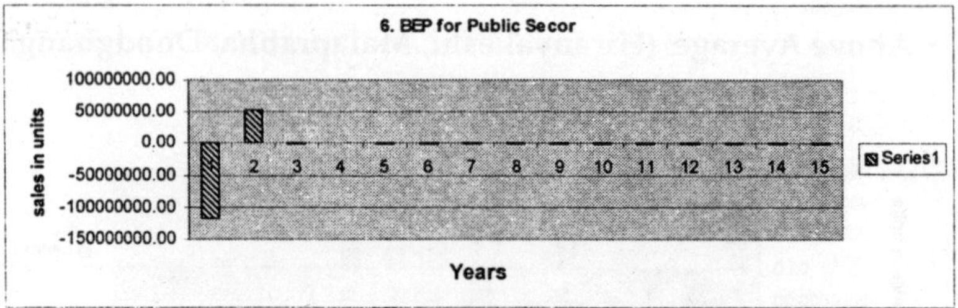

It is found that public sector sugar factories have seen a negative BEP in the beginning and some how achieved positive BEP sales during 1992-93 and afterwards public sector has no indication of BEP sales.

III. Private Sector

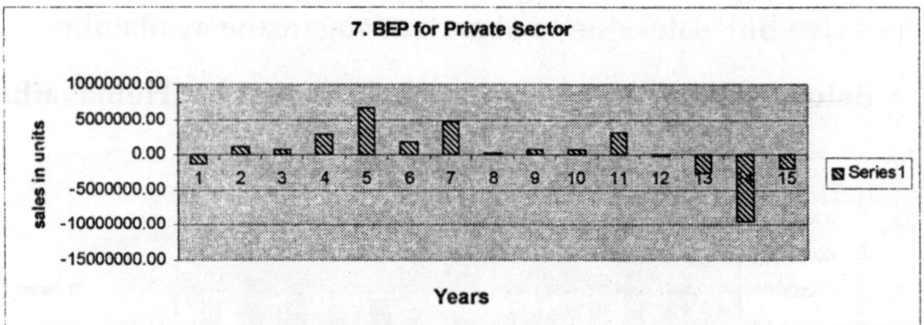

It is found that private sector sugar factories were on safer side in the beginning but for the last 4 years of time this sector is also facing problem, as its variable cost of production is more than selling price. So it has achieved Negative BEP from the last four years.

5.3.D. Safety Margin

Safety Margin is a management control technique which emphasises to decrease or increase the production of sugar to achieve breakeven sales at which sugar factory earns no profit/ no loss. Public Sector is more predominant in achieving Safety Margin, which is expressed in percentage of breakeven sales.

Safety Margin is positive in the beginning but the latter the years Safety Margin has comedown.

The following are the analyses made for safety margin and presented in the Tables 5.10, 5.10a and Graphs 8 to 10.

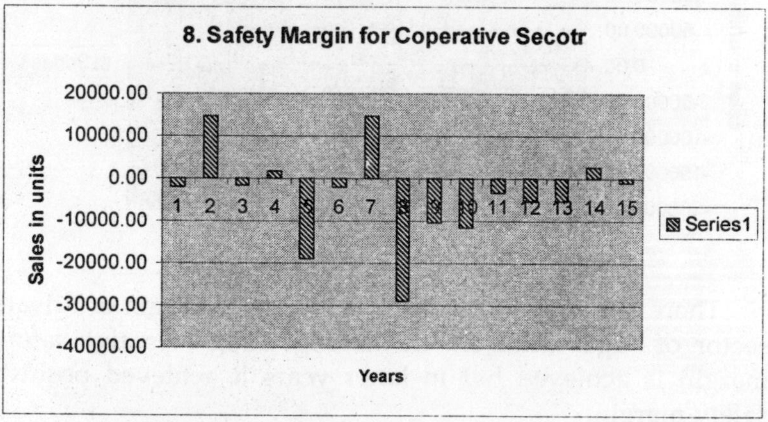

It is found in case of co-operative sector at least 3000 quintals of surplus sugar to be produced with out altering fixed cost to achieve safety margin or to achieve BEP sales.

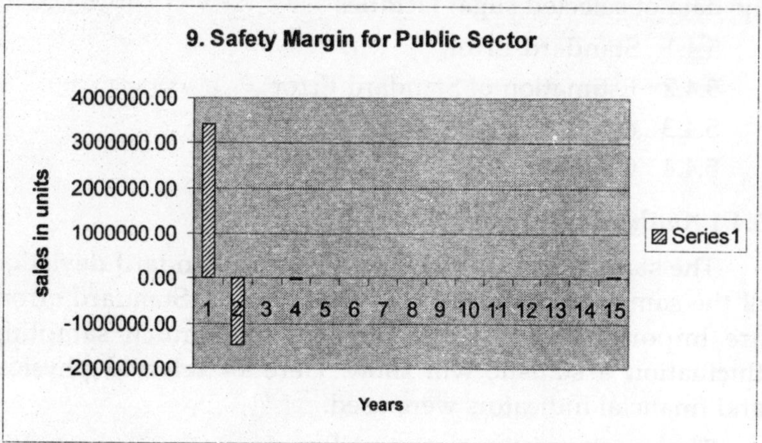

It is observed that public sector is more predominant in achieving safety margin in the beginning but latter the years Safety Margin is less.

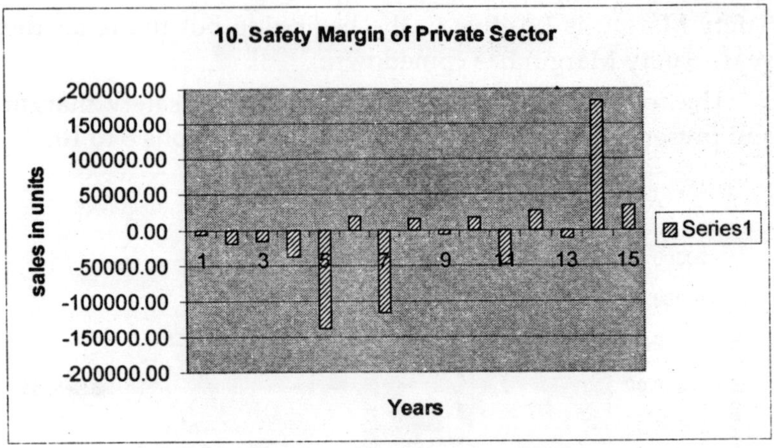

There is fluctuation in achieving safety margin of Private sector of sugar industry. In the beginning Negative safety margin is achieved but in latter years it achieved positive safety margin.

5.4 Statistical Analyses of the Sugar Factories

The following are the statistical tools used for analysing the data of selected sugar factories

5.4.1 Standard Error

5.4.2 Estimation of Standard Error

5.4.3 t-Test

5.4.4 Compound Growth Rate (CGR)

5.4.1 *Analyses of standard error*

The standard error of a statistic is the standard deviation of the sampling distribution of that statistic. Standard errors are important because they reflect how much sampling fluctuation a statistic will show. Here 14 selected physical and financial indicators were used.

The results of the statistical l analyses are presented in the Tables 5.11 and 5.11a, Graphs 11 to 24. and Appendix 44.

Table 5.11

STANDARD ERROR OF ALL THE PHYSICAL AND FINANCIAL INDICATORS FROM 1990-91 TO 2004-05

Year/ Variables	*Sugar Production (Quintals)*	*Quantity of Sugarcane Crushed (Tones/Year)*	*Recovery Percentage*	*Capacity Utilisation (Tones/Day)*	*Share Capital (Lacks)*	*Net Profit (Lacks)*	*Barrowings (Lacks)*
1990 - 91	318858.15	288354.605	0.548	1368.397	466.353	87.648	990.91
1991 - 92	326874.331	258754.898	1.101	1278.269	431.898	124.723	908.149
1992 - 93	337024.94	214612.01	1.302	1393.595	481.396	182.47	865.595
1993 - 94	287072.912	210352.515	1.451	1385.107	710.386	172.662	1203.866
1994 - 95	402945.084	322480.471	1.188	1398.402	699.036	268.396	1589.322
1995 - 96	440140.098	366712.84	0.845	1350.464	694.375	318.422	2288.849
1996 - 97	371325.643	278617.133	1.358	1419.487	688.705	218.967	1778.109
1997 - 98	429921.094	360247.271	3.489	1796.17	672.991	182.566	2196.839
1998 - 99	469188.327	404014.305	1.516	1613.347	660.682	962.187	3476.492
1999 - 00	516736.67	418303.951	1.2	1725.656	778.176	711.807	3530.276
2000 - 01	457703.899	355546.956	1.148	1687.607	773.841	800.665	3636.191
2001 - 02	472237.12	342162.01	1.10	1873.50	788.90	686.70	4200.57
2002 - 03	479986.52	379676.59	1.23	1976.73	789.56	985.33	4621.32
2003 - 04	256057.29	211986.19	3.21	1892.75	752.84	652.50	5243.06
2004 - 05	538769.31	457044.31	3.49	2415.83	1000.91	678.12	6594.60

Source: Compiled and Analysed from the annual reports and general ledgers of sample sugar factories for the above mentioned physical and financial parameters from 1990-91 to 2004-05.

Table 5.11a

STANDARD ERROR OF ALL THE PHYSICAL AND FINANCIAL INDICATORS FROM 1990-91 TO 2004-05

Year/ Variables	*Current Liabilities (Lacks)*	*Level of Reserve Funds (Lacks)*	*Total Expenses (Lacks)*	*Gross Income (Lacks)*	*Total Assets Position (Lacks)*	*Total Sales (Lacks)*	*Fixed Assets Position (Lacks)*
1990 - 91	952.32	574.018	3167.954	3200.458	193085.55	2144.877	2527.135
1991 - 92	1084.663	529.055	3606.657	3734.445	2905.193	2009.568	1980.34
1992 - 93	1621.217	436.828	4398.92	4561.101	3183.655	2517.641	1833.262
1993 - 94	1930.073	368.951	5528.039	5712.652	4256.488	3293.259	1656.676
1994 - 95	1496.481	438.485	7046.209	7184.998	4102.887	5365.803	1750.422
1995 - 96	1504.887	519.204	7516.017	7715.077	4819.106	4653.293	2822.643
1996 - 97	1778.469	578.795	7515.429	7742.846	5087.014	4906.989	2976.772
1997 - 98	1720.034	598.898	8791.666	8867.434	5853.059	5412.919	3704.376
1998 - 99	1288.93	723.092	9569.569	10137.33	6558.355	5286.965	3665.077
1999 - 00	1955.133	895.935	10403.431	11139.69	7467.014	6258.844	4211.558
2000 - 01	1681.31	1263.467	10257.571	31370.89	7974.013	6438.583	2635.452
2001 - 02	1574.86	1159.50	10992.23	11684.7	8721.565	6286.31	2677.155
2002 - 03	1906.40	1108.09	10905.38	11667.65	8943.516	6073.06	3259.068
2003 - 04	2276.42	1039.22	9195.39	9958.306	9275.174	5795.383	22855.543
2004 - 05	3418.45	1215.02	7860.61	8590.222	9662.655	6495.57	3886.483

Source: Compiled and Analysed from the annual reports and general ledgers of sample sugar factories for the above mentioned physical and financial parameters from 1990-91 to 2004-05

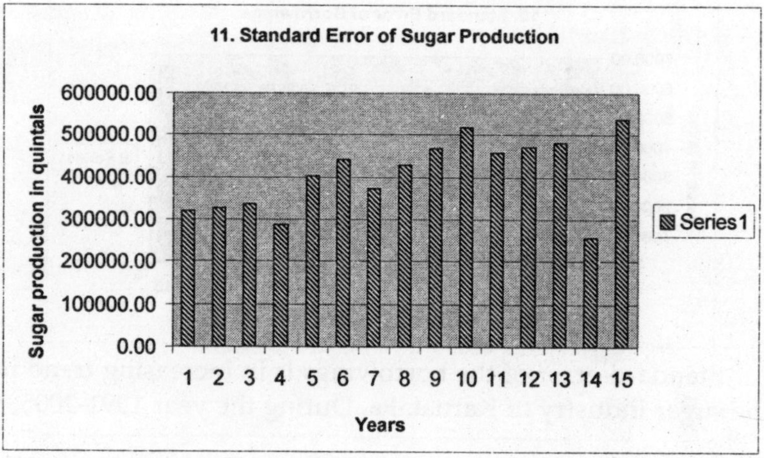

It is found that standard error of sugar production of overall sugar industry is highest in the year 2004-05 and followed by the year 1999-00 and 2002-03.

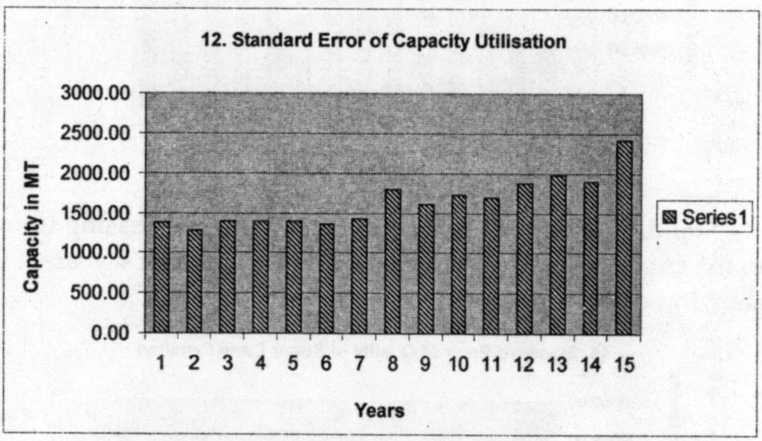

Standard error of the capacity utilisation is in increasing trend in the sugar industry of Karnataka.

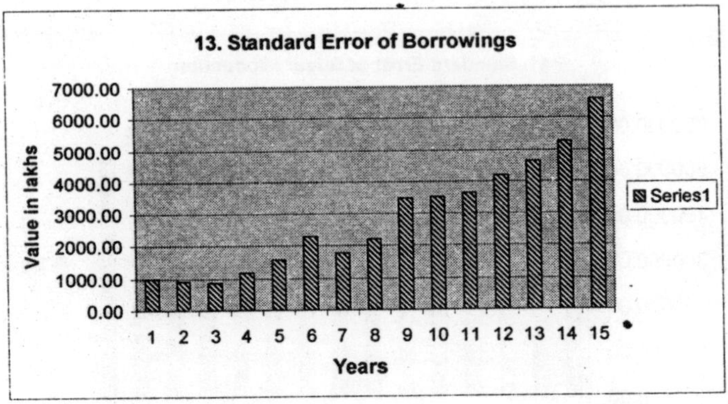

Standard error of the borrowings is in increasing trend in the sugar industry of Karnataka. During the year 1991-2005.

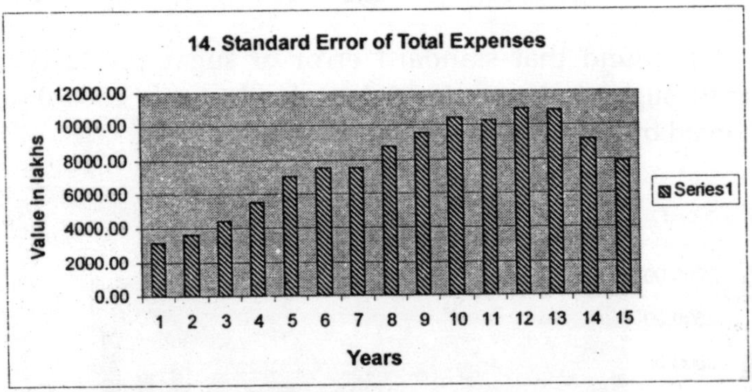

Standard error of the total expenses is in increasing trend in the sugar industry of Karnataka. But in the last 4 years it is showing declining trend.

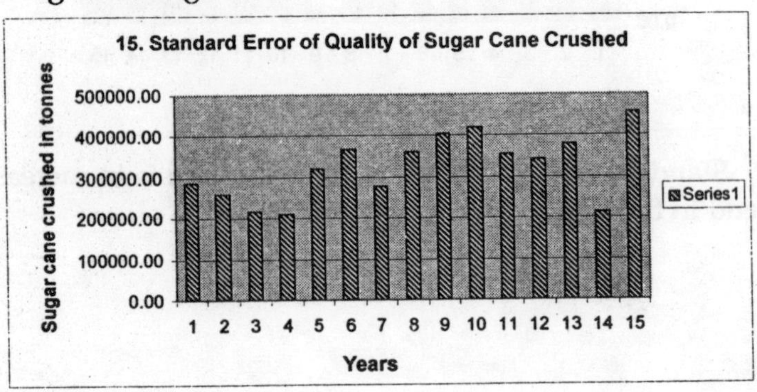

Standard error of the quantity of sugarcane crushed is varied from year to year in 15 years of time because of non-availability of sugarcane in the field in sufficient quantity.

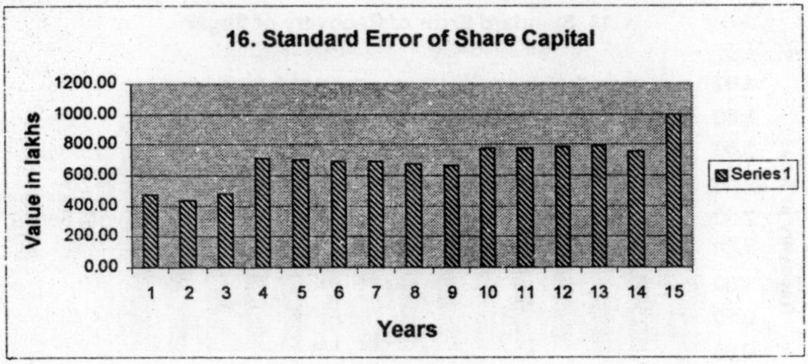

Standard error of the share capital is increased moderately from year to year during the study period.

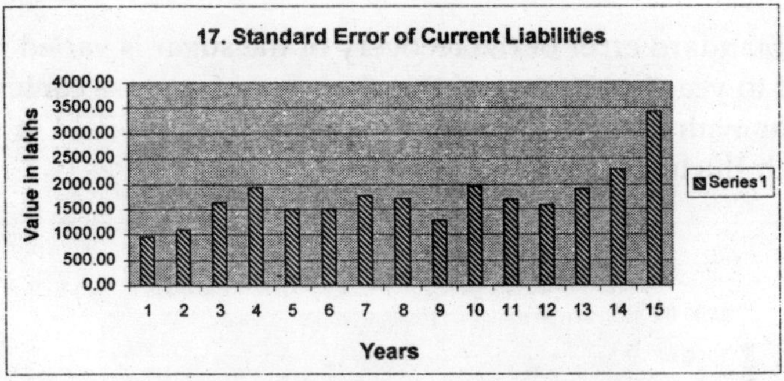

Standard error of the current liabilities increased from year to year in 15 years of time because of borrowings were increased from year to year.

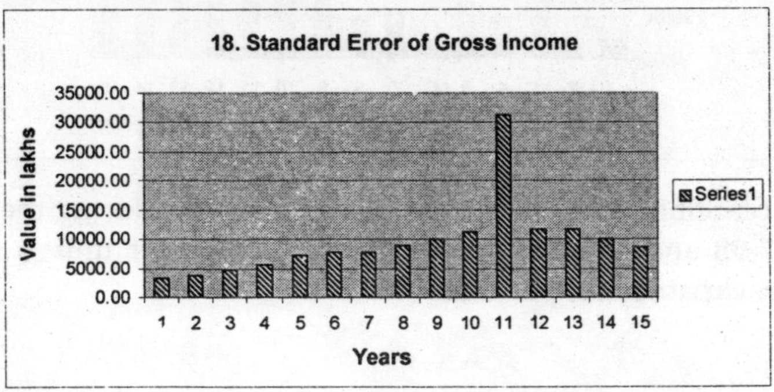

Standard error of the gross income was varied from year to year and it was high during 2000-01.

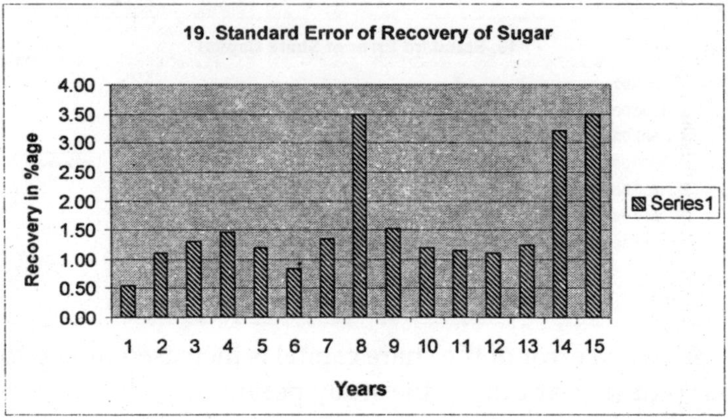

Standard error of the recovery of the sugar is varied from year to year in 15 years of time because of sucrose content of sugar with respect to yield of sugarcane in the field is very much fluctuated.

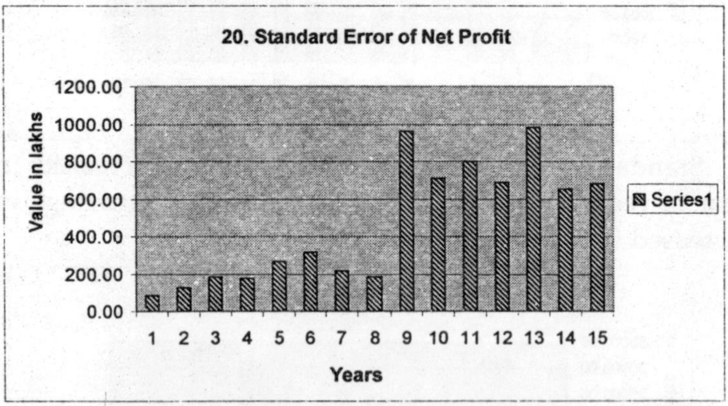

Standard error of the net profit increased during the year 1997-98 and 2002-03 because selling price per unit is more than variable cost per unit.

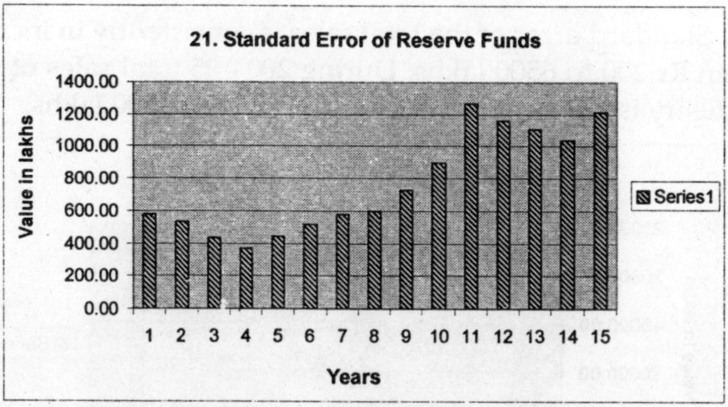

Standard error of the reserve funds varied from year to year in 15 years of time. During 2004-05 it is following positive trend.

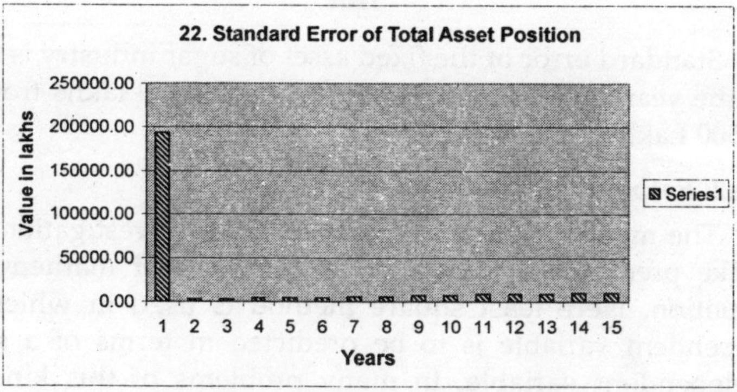

Standard error of the total asset position is more in the beginning and further is reduced over 14 years of time.

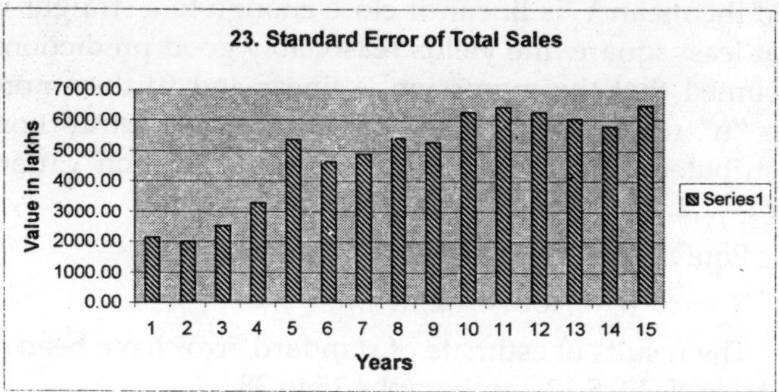

Standard error of the total sales is consistently in increases from Rs 200 to 6500 lakhs. During 2004-05 total sales of sugar industry is increased from Rs 5750 lakhs to 6500 lakhs.

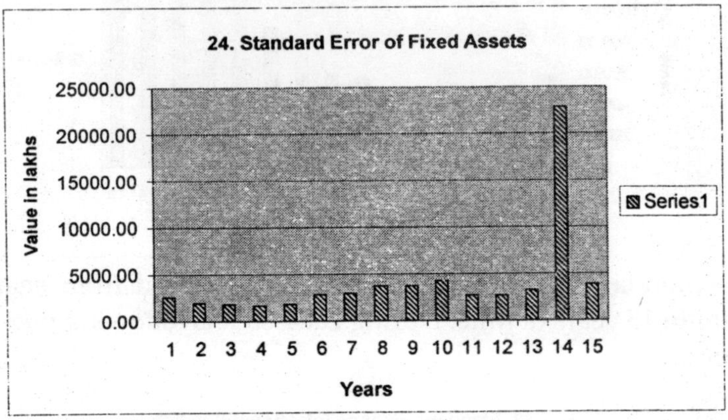

Standard error of the fixed asset of sugar industry is more in the year 2003-04. and is reduced to Rs 250 lakhs from Rs 22500 Lakhs.

5.4.2. *Estimate of standard error*

The main objective of much statistical investigation is to make predictions preferably on the basis of mathematical equation. Here least square method is used in which the dependent variable is to be predicted in terms of a single independent variable. In many problems of this kind the independent variable is observed without error or with the error in the dependent variable. The relationship between X and the mean Y_i is linear or close enough to a straight-line so that least square line yields reasonably good predictions. It is assumed that the regression is linear and further more that the "n" random variables "Y_i" are independently normally distributed with the means $\alpha + \beta_i$ and common variance σ^2 for i = 1,2,......n.

Equivalently we write the equation as

$Y_i = \alpha + \beta_i + e_i$ for i = 1,2,......n.

The results of estimate of standard error have been shown in tables 5.12, 5.12a and graphs 25 to 38.

TABLE 5.12

ESTIMATION OF STANDARD ERROR

Sl No	Selected Factories	Sugar Production	Quantity of Sugarcane Crushed	Recovery Percentage	Capacity Utilisation	Share Capital	Net Profit	Barrowings
1	Hiranyakeshi	25.01	24.25	5.716	6.48	48.62	48.62	18.75
2	Malaprabha	22.14	21.77	2.74	6.62	24.45	24.45	40.62
3	Doodghanga	26.81	21.99	5.32	9.63	19.57	19.57	21.81
4	Pandavapura	52.57	54.19	42.17	51.44	24.09	24.09	65.81
5	Ghataprabha	20.46	19.65	3.31	25.05	5.10	5.10	11.74
6	Hemavathi	37.74	49.50	6.510	35.26	8.42	8.42	38.60
7	My Sugar	113.34	27.84	10.53	17.54	17.45	17.45	27.26
8	Mysore	15.25	19.30	3.49	9.88	8.25	8.25	13.98
9.	Ugar	19.76	20.10	5.03	7.43	13.17	13.17	15.93
10.	Dhavanagere	34.62	31.61	6.16	22.32	67.91	67.91	42.59

***Source:** Compiled and Analysed from the annual reports and general ledgers of sample sugar factories for the above mentioned physical and financial parameters from 1990-91 to 2004-05*

TABLE 5.12a

ESTIMATION OF STANDARD ERROR

Sl No	*Selected Factories*	*Current Liabilities*	*Level of Reserve Funds*	*Total Expenses*	*Gross Income*	*Total Assets Position*	*Total Sales*	*Fixed Assets Position*
1	Hiranyakeshi	27.14318	6.077624	33.73955	39.95	20.08107	17.12	5.68
2	Malaprabha	28.48124	13.3119	30.11222	106.09	26.54343	24.79	4.430
3	Doodghanga	45.40253	14.03717	37.26384	23.39	5.960882	14.60	33.00
4	Pandavapura	79.41582	50.5764	90.63747	191.65	64.24763	64.09	78.55
5	Ghataprabha	32.42102	10.90144	16.63905	37.589	10.68128	19.38	11.81
6	Hemavathi	31.338	8.254613	24.64602	131.26	32.28957	13.88	17.48
7	My Sugar	47.26301	24.21375	33.70868	37.925	62.5296	34.50	191.68
8	Mysore	25.20138	41.38214	12.03713	178.66	288.3338	27.96	34.30
9.	Ugar	13.20778	25.08434	12.46258	12.463	13.3912	15.54	20.25
10.	Dhavanagere	41.051	60.37348	38.55992	29.76	39.06081	42.32	78.92

Source: *Compiled and Analysed from the annual reports and general ledgers of sample sugar factories for the above mentioned physical and financial parameters from 1990-91 to 2004-05*

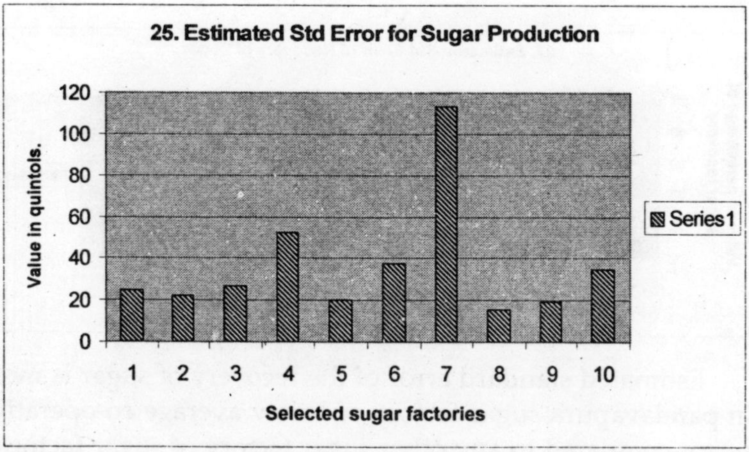

Estimated standard error of the sugar production is more in My Sugar factory of public sector compared to other co-operative and private sectors and estimated standard error of sugar production of private sector is low compared to co-operative sugar factories having the estimated value ranging from 15.25 per cent to 113.34 per cent.

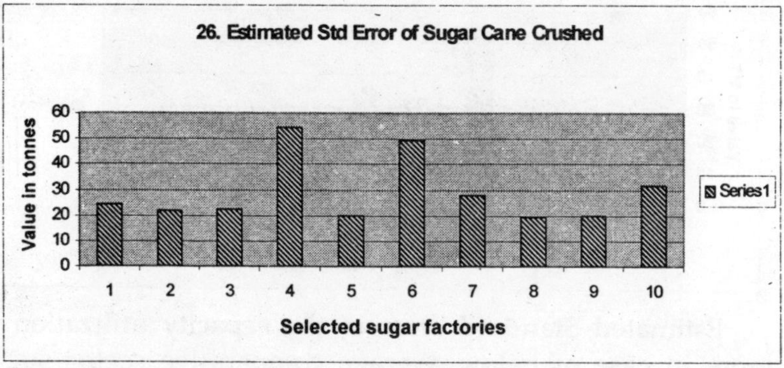

Estimated standard error of the sugarcane crushed is reasonably more in case of below average co-operative sector than the above average co-operative and private sugar factories having the value ranging from 19 to 50 per cent of estimated standard error.

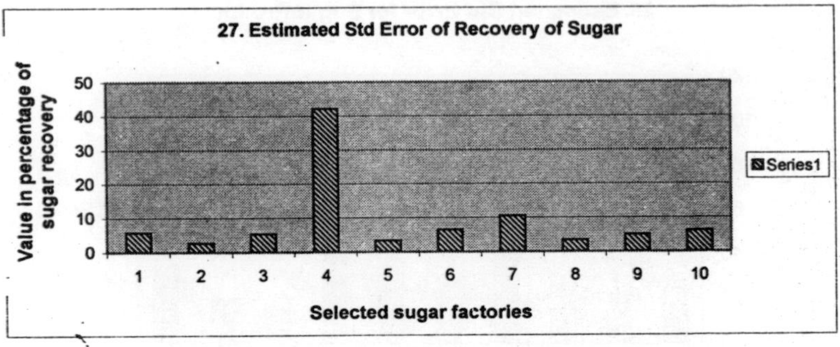

Estimated standard error of the recovery of sugar is more in pandavapura sugar factory of below average co-operative sector compared to any other sugar factory of sugar factories in Karnataka. This is due to various factors suchas company has not adopting diffuser mechanism which increases the yield adopting of sugar recovery. The estimated error expressed in percentage is ranging from 5 to 42 per cent in Karnataka.

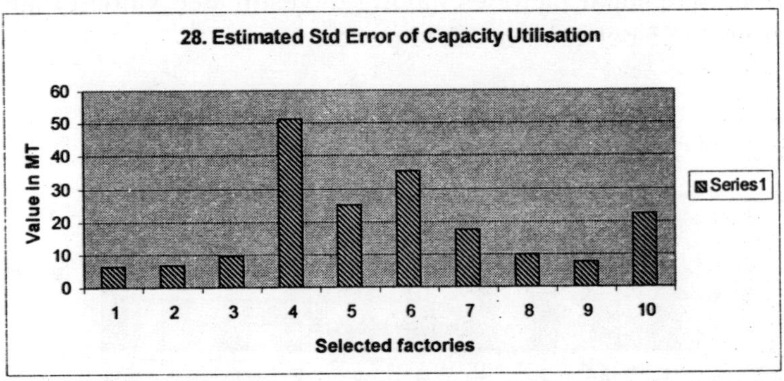

Estimated Standard error of the capacity utilization is more in case of below average co-operative sector sugar factories that includes Pandavapura, Ghataprabha and Hemavathi sugar factories. But the estimated error of capacity utilization of the private sector compared to the co-operatives of above average sugar factories appears to be more productive.

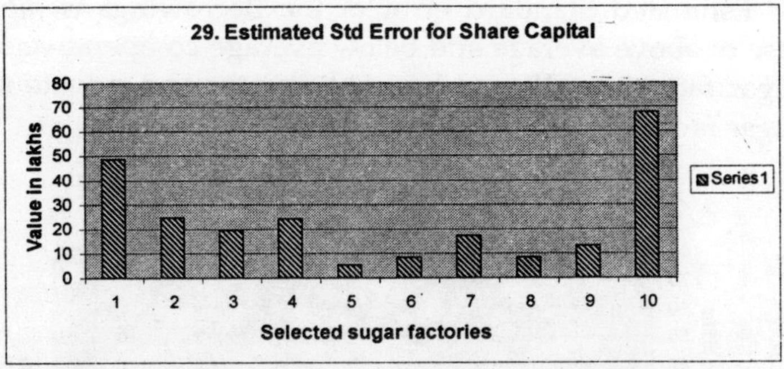

Estimated standard error of the share capital is more in case of private sugar factory i.e., Dhavanagere sugar factory followed by Hiranyakeshi sugar factory of above average co-operative sector.

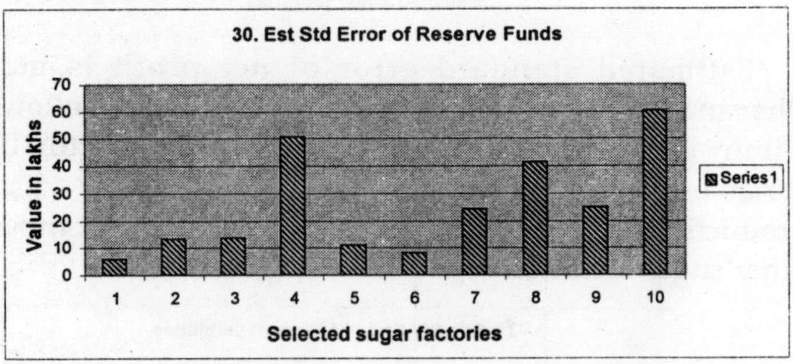

Estimated standard error of the reserve funds is more in private sector compared to public sector and followed by below average sugar factory i.e. Pandavapura sugar factory. Incase of public sector it is low.

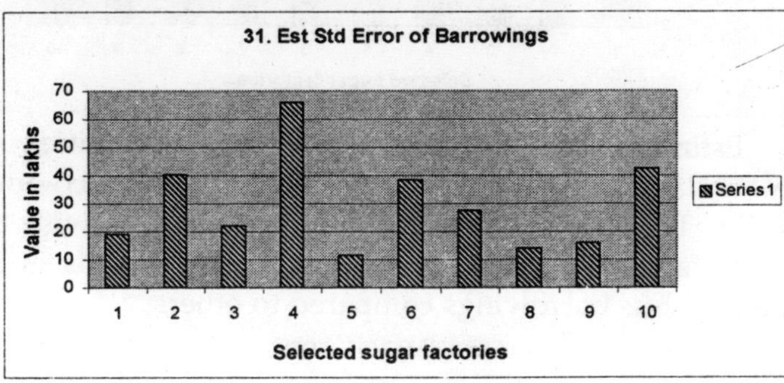

Estimated Standard error of the Borrowings is more in case of above average and below average co-operative sector sugar factories followed by public sector and private sector sugar factories.

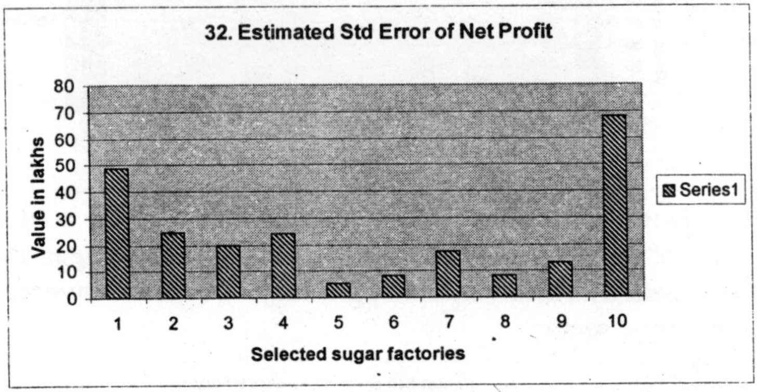

Estimated standard error of net profit is more in Dhavanagere sugar factory of private sector and followed by Hiranyakeshi sugar factory of co-operative sector due to variable cost is more than the selling price of the sugar and production of sugar in these two factories is less compared to other sugar factories.

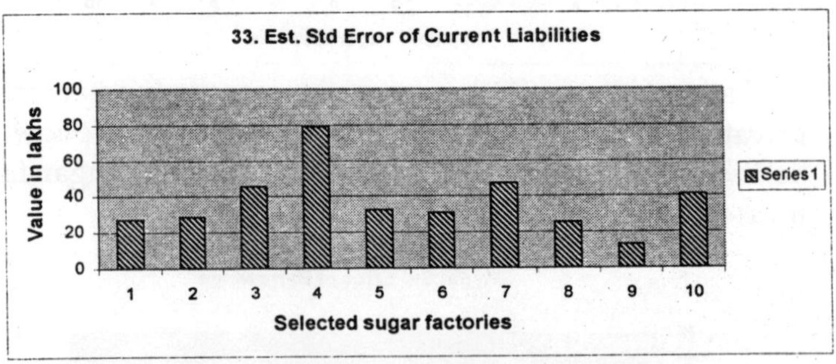

Estimated standard error of the current liabilities is more in Pandavapura sugar factory of below average co-operative sector. The lowest estimated standard error is with private sugar factory i.e. Ugarkhurd sugar factory. This is due to their for less borrowings compared to others.

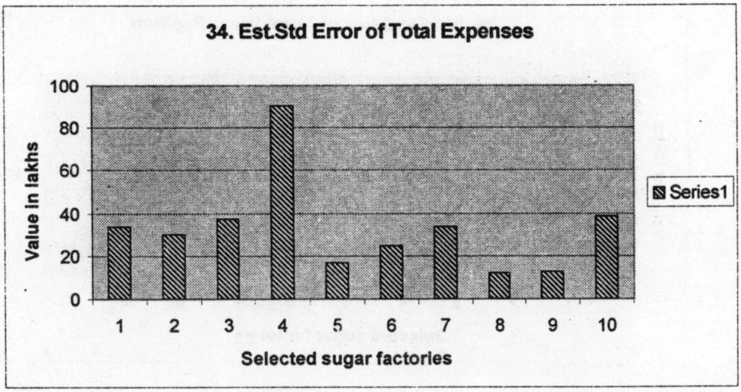

Estimated standard error of the total expenses is more in Pandavapura sugar factory of below average co-operative sector and followed by above average co-operative sugar factories. The lowest estimated standard error is with Ugarkhurd sugar factory of private sector in Karnataka.

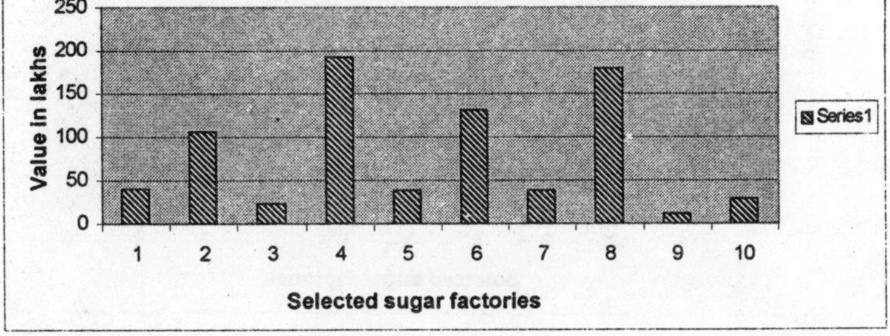

Estimated standard error of the gross income is less incase of private sugar factories such as Ugarkhurd and Davanagere sugar factory and followed by public sector sugar factories and co-operative sugar factories.

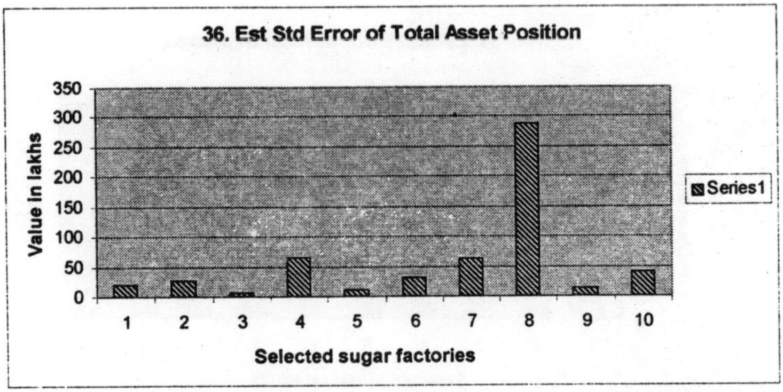

Estimated standard error of the total asset position is more in Mysore paper mills sugar factory, Bhadravathi. The estimated standard error of total asset position of Doodhganga sugar factory is more followed by Ghataprabha sugar factory and Ugarkhurd sugar factory of private sector.

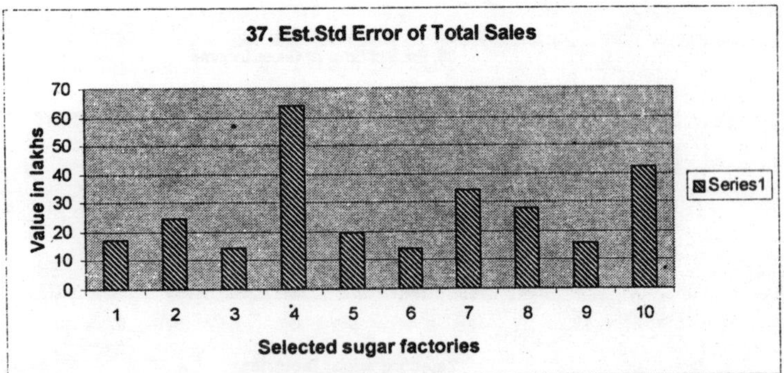

Estimated standard error of the total sales is more in Pandavapura sugar factory of below average co-operative sector and followed by Dhavangere Sugar factory of private sector.

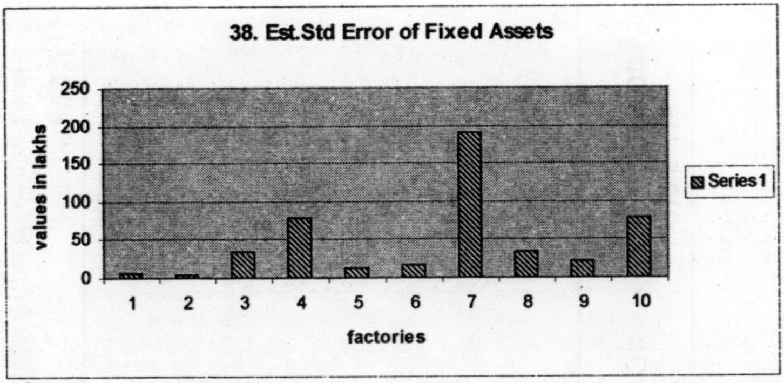

Estimated standard error of the fixed asset position is more in case of My sugar factory of public sector and followed by Pandavapura sugar factory of below average co-operative sector.

5.4.3. *t-test*

t-test is a test of hypothesis used to determine the significant difference between sample mean and assumed mean by holding tentative statements which to be tested by the researcher with the degree of freedom. As t-value calculated and computed in the table for various selected physical and financial indicators of the project are considered to be significant difference in their mean values if t- statistic value is $|t|<$ table value (2.145) at 5 per cent level of significance, If computed t – values are less than table value, then the difference is not significant other wise it is significant. Level of reserve funds and total liabilities of public sector have shown significant difference from its mean value as table value is more than calculated t-value i.e. Appendix-.11 (Level of Reserve Funds), and Appendix-18 (Total Liabilities).

The results of the t-test are shown in Table 5.13, Graph 39 and Appendix 45.

TABLE: 5.13

t-TEST FOR AVERAGE SELECTED INDICATORS

SL NO.	INDICATORS	OVER ALL CO-OPERATIVE SECTOR AVG	OVER ALL PUBLIC SECTOR S.F AVG	OVER ALL PRIVATE SECTOR S.F AVG
1	QUANTITY OF SUGARCANE CRUSHED	* -2.018	* -1.65	* -6.83
2	SUGAR PRODUCTION	* -1.63	* -3.30	* -4.17
3	SHARE CAPITAL POSITION	* -7.45	* -4.539	* -4.01
4	LEVEL OF RESERVE FUNDS	* -5.038	** 4.29	* -4.48
5	LEVEL OF OWNED FUNDS	* -6.532	* 0.07	* -4.53
6	CURRENT ASSETS POSITION	* -7.699	* -1.39	* -2.80
7	FIXED ASSETS POSITION	* -2.41	* 0.40	* -2.87
8	TOTAL ASSETS POSITION	* -0.674	* 0.83	* -4.47
9	CURRENT LIABILITIES	* -2.766	* -3.91	* 0.34
10	TOTAL BORROWINGS	* -3.374	* -1.717	* -3.25
11	TOTAL LIABILITIES	* -4.35	** 2.22	* -3.65
12	TOTAL SALES	* -4.016	* -4.96	* -3.51
13	GROSS INCOME	* -1.861	* -0.38	* -3.73
14	TOTAL EXPENSES	* -2.308	* -2.75	* -2.97

* Not significant, ** significant (R: $|t| \leq 2.145$)

Source: *Compiled and Analysed from the annual reports and general ledgers of sample sugar factories for the above mentioned physical and financial parameters from 1990-91 to 2004-05.*

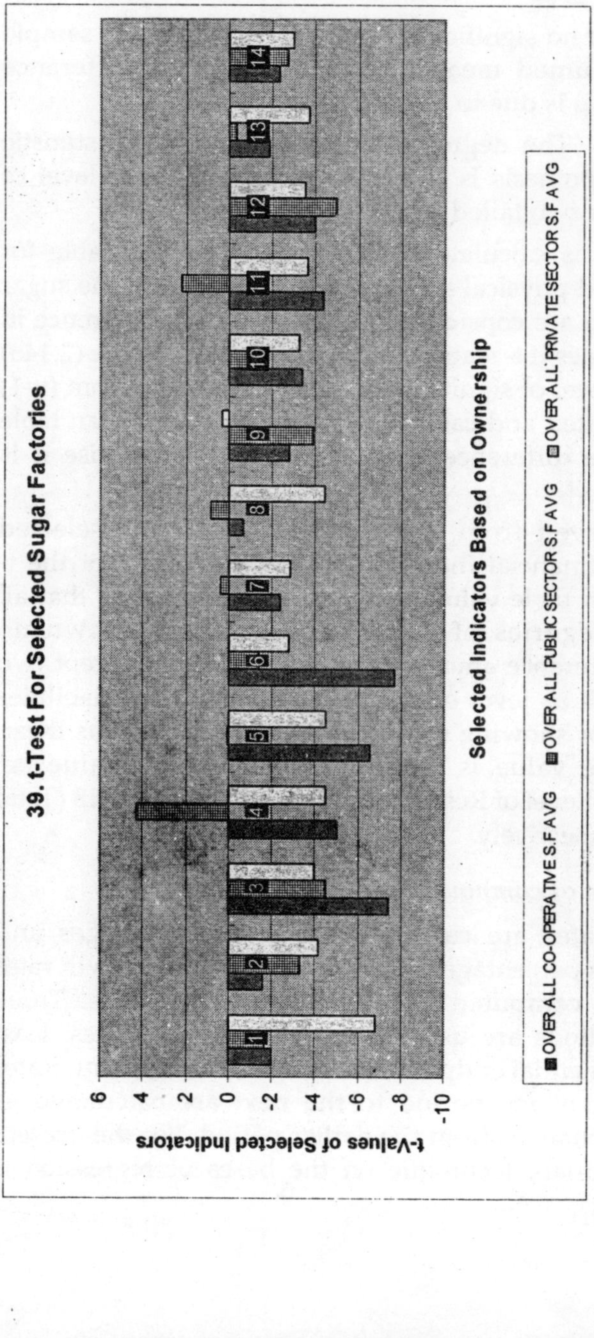

1 QUANTITY OF SUGARCANE CRUSHED
2 SUGAR PRODUCTION
3 SHARE CAPITAL POSITION
4 LEVEL OF RESERVE FUNDS
5 LEVEL OF OWNED FUNDS

6 CURRENT ASSETS POSITION
7 FIXED ASSETS POSITION
8 TOTAL ASSETS POSITION
9 CURRENT LIABILITIES
10 TOTAL BORROWINGS

11 TOTAL LIABILITIES
12 TOTAL SALES
13 GROSS INCOME
14 TOTAL EXPENSES

Here the null hypothesis is set up as H_0: $\mu = \mu_0$ i.e. the assumed mean is μ_0

$H_{0:}$ there is no significant difference between the sample mean and assumed mean. In other words the difference between x and μ is due to fluctuations of sampling

$H_{a:}$ $\mu \neq \mu_0$. The degree of freedom for above t-statistic under null hypothesis is (n-1) i.e., 14 at 5 per cent level of significance for two tailed test..

As t-value is calculated and computed in the table for various selected physical and financial indicators of the sugar factories which are considered to be significant difference in their mean values if t- statistic value is $|t|$ < table value (2.145) at 5 per cent level of significance and degree of freedom (n-1) is 14. If computed and calculated t-values are less than table value then the difference is not significant, otherwise it is significant.

It is observed from above table that all the selected physical and financial indicators (variables) have got the t-value less than table value (2.145) so it is concluded that all the three categories of sugar factories have shown no significant difference among the listed indicators except two indicators such as level of reserve funds and total liabilities of public sector showing significant difference from its mean value as table value is more than calculated t-value i.e. Appendix-11 (Level of Reserve Funds) and Appendix-18 (Total Liabilities) respectively.

5.4.4. *Analyses of compound growth rate (CGR)*

Growth rates are calculated as annual averages and represented as percentages. Except where noted, growth rates of values are computed from constant price series. Three principal methods are used to calculate growth rates: least squares, exponential endpoint, and geometric endpoint. Rates of change from one period to the next are calculated as proportional changes from the earlier period. For the present study least square technique on the bases of regression is applied

The results of the CGR are shown in Table 5.14 and Graph 40.

TABLE: 5.14

CGR FOR SELECTED INDICATORS

Sl No	*Indicators*	*Co-operative*	*Public*	*Private*
1	INSTALLED CAPACITY	2.28	0.00	4.54
2	SUGARCANE CRUSHED	-0.72	1.14	3.25
3	RECOVERY PERCENTAGE	10.10	9.42	10.72
4	SUGAR PRODUCTION	-0.78	-2.06	2.99
5	SHARE CAPITAL POSITION	5.60	5.05	15.05
6	LEVEL OF RESERVE FUNDS	5.83	-4.73	15.59
7	LEVEL OF OWNED FUNDS	5.71	1.70	15.40
8	CURRENT ASSETS POSITION	9.65	6.31	9.39
9	FIXED ASSETS POSITION	7.77	14.20	15.79
10	TOTAL ASSETS POSITION	9.84	-28.35	11.62
11	CURRENT LIABILITIES	3.55	12.61	6.08
12	TOTAL BORROWINGS	12.60	11.84	14.47
13	TOTAL LIABILITIES	9.77	12.18	11.80
14	TOTAL SALES	5.82	6.26	7.13
15	GROSS INCOME	5.80	6.51	8.05
16	TOTAL EXPENSES	5.29	5.18	8.07
17	NET PROFIT	-49.28	-92.00	134.36

Source: Compiled and Analysed from the annual reports and general ledgers of sample sugar factories for the above mentioned physical and financial parameters from 1990-91 to 2004-05

Compound growth rate of installed capacity of private sector of sugar factories was 4 per cent over 15 years of time and followed by co-operative sector of sugar factories that was about 2.5 per cent. This increased compound growth rate of installed capacity of private sector was accounted for increase in the sugar production. Compound growth rate of installed capacity of public sector was nil. It means that growth has become stagnant.

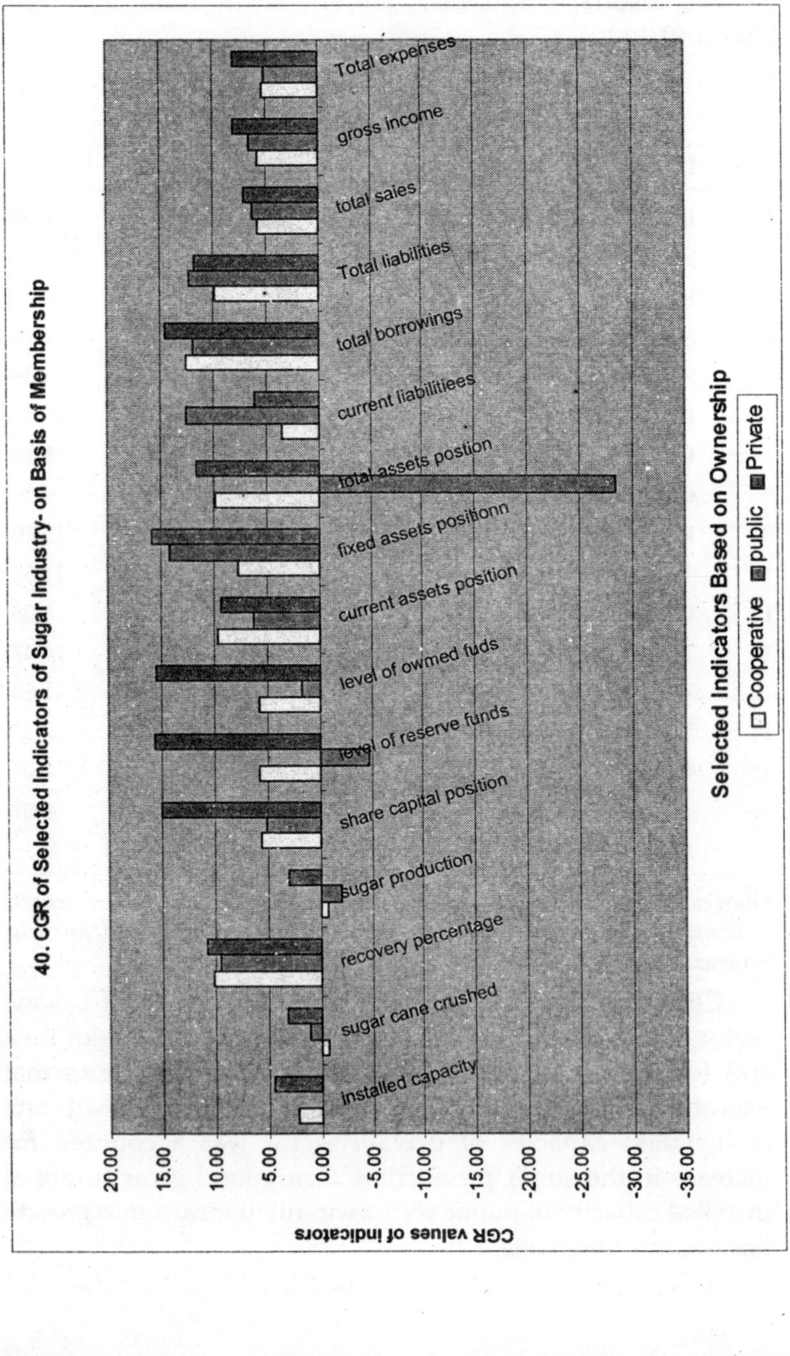
40. CGR of Selected Indicators of Sugar Industry- on Basis of Membership
Total expenses
gross income
total sales
Total liabilities
total borrowings
current liabilitiees
total assets postion
fixed assets positionn
current assets position
level of owmed fuds
level of reserve funds
share capital position
sugar production
recovery percentage
sugar cane crushed
installed capacity
20.00
15.00
10.00
5.00
0.00
-5.00
-10.00
-15.00
-20.00
-25.00
-30.00
-35.00
CGR values of indicators
Selected Indicators Based on Ownership
Cooperative
public
Private

Compound growth of sugar cane crushed incase of co-operative sector sugar factories was less than zero although the installed capacity's compound growth rate was about 2.5 per cent this was because number of crushing days are less compared to public sector and private sector. In case of private sector sugar factories cane crushing was more and followed by public and private sector sugar factories.

Compound growth rate of recovery of sugar was more in private sector as compared to co-operative and public sector sugar factories. Private sugar factories have diffuser mechanism system (DMS), which enhances the recovery of sugar from the sugarcane. Private sector was accounted for more than 10 per cent of recovery of sugar & followed by co-operative sector accounted for 10 per cent and co-operative sector accounted for below 10 per cent.

Compound growth rate of sugar production was more incase of private sector the only sector accounted for just 2.5 per cent growth compared to other co-operative and public sector which accounted for below zero per cent of growth in the sugar production. The very less sugar production was observed in case of public sector sugar factories.

Compound growth rate of share capital was 15 per cent and followed by co-operative and public sector accounted for 6 per cent and 5 per cent respectively

Compound growth rate of level of reserve funds was more in private sector accounting for 15 per cent growth and followed by co-operative sector accounting for 6 per cent growth. Incase of public sector compound growth rate was negative that was 5 per cent during the study period.

Compound Growth Rate of level of owned funds was more incase of private sector accounting for 16 per cent growth and followed by co-operative sector resulting in 6 per cent growth and public sector resulting in 2 per cent growth

Compound Growth Rate of level of current asset position was more incase of co-operative sector of sugar factories accounting for 9 per cent growth and followed by private sector accounting for 8 per cent growth. But incase of public sector growth rate was just about 6 per cent this was the weakness of the sector in sugar factories.

Compound Growth Rate of level of fixed asset position was more in case of private sector accounting for 16 per cent growth and followed by 14 per cent growth rate of public sector and 6 per cent growth rate of co-operative sector of sugar factories.

Compound Growth Rate of level of total asset position was determined to be more in case of private sector as its CGR of fixed asset position was more and although its current asset position was less comparatively. And followed by co-operative sector accounting for 10 per cent growth. Incase of public sector the compound growth rate was negative resulting in 23 per cent growths.

Compound Growth Rate of level of current liabilities was more in case of public sector of sugar factories accounting for 15 per cent growth and followed by private sector accounting for 6 per cent growth and co-operative sector resulting in 3 per cent growth

Compound Growth Rate of level of total borrowings was more in case of private sector accounting for 14 per cent growth and followed by 13 per cent growth of co-operative sector and 12 per cent growth of public sector. Here it was observed that total sugar factories were set equally free to borrow, so borrowings' compound growth rate was by large equal with respect to membership of the sugar factories.

Compound Growth Rate of level of total liabilities was less in case of co-operative sector accounting for about 10 per cent growth as its current liabilities were very less compared to other sector. But incase of both private and public sector of sugar factories the compound growth rate of level of total liabilities was more or less same accounting for 12-13 per cent growth.

Compound Growth Rate of level of total sales was more in case of private sector accounting for 8 per cent growth and followed by 7 per cent growth of public sector and 6 per cent growth of co-operative sector of sugar factories.

Compound Growth Rate of level of Gross income was more in case of private sector accounted for 9 per cent growth and followed by 8 per cent growth of public sector and 7 per cent growth of co-operative sector of sugar factories. Here it was observed that the change in the compound growth rate of total sales of sugar and the change in the compound growth rate of gross income are in same proportion with respect to different ownership of the sugar factories.

Compound Growth Rate of level of total expenses was more in case of private sector accounting for 8 per cent growth and followed by both public and co-operative sector accounting for 5 per cent growth. This was because private sector meets the additional expenses of additional units, which are supportive in enhancing recovery of sugar. And in private sector compound growth rate of level of fixed asset position was more compared to co-operative and public sector of sugar factories as explained earlier.

Compound Growth Rate of level of net Profit was more in case of private sector compared to co-operative sector and public sector of sugar factories in Karnataka. The compound growth rate of net profit of co-operative sector and public sector have shown negative trend. That means both co-operative and public sector of sugar factories were suffering more than the private sector did.

5.5 Capacity Utilization of Sugar Factories

The capacity utilization of sugar factory depends mainly on the installed capacity and availability of cane for crushing.

Over the study period the overall average capacity utilization was to the extent of 69.48 per cent (Table 5.15, 5.15a and 5.16).

The capacity utilization ranged between 20.24 per cent in Pandavapura (below average co-operative) to 98.78 in the Gataprabha (below average co-operative) sugar factory, where as the lowest and the highest of capacity utilization were

67.33 per cent in My Sugar (above average public) and 95.95 per cent in Mysore paper mills (below average public). It was ranged from 59.33 to 77.83 per cent in private sector.

The capacity utilization was the highest in Ghataprabha co-operative sugar factory (98.78 per cent) and the lowest in Pandavapura co-operative sugar factory (20.14 per cent). The overall average capacity utilization of co-operative sugar factories, public sector sugar factories and private sector sugar factories were 55.05, 88.79 and 64.34 per cent respectively. The average capacity utilization of the Pandavapura co-operative sugar factory was the lowest because it was not working during 1996-97, 2002-03 and 2004-05.

The above average sugar factories had higher capacity utilization than the below average sugar factories in the cases of co-operative and public sectors, where as it was vice-versa in the case of private sector sugar factories.

5.6 Cost of Processing Sugarcane Into Sugar

The cost of processing sugarcane into sugar in different factories is presented in Table 5.17. The cost of processing was worked out for the year 2004-05 for all the factories except Pandavapura co-operative sugar factory (for this factory cost of processing was calculated for the year 2002-03) for the reasons quoted under sub-head 5.5.

On an average the cost of cane accounted for 51.88 per cent of the total cost of processing, followed by fixed costs (28.16 per cent) and other variable costs (23.75 per cent).

With respect to the cost of sugarcane, it was the highest in private sector sugar factories (Rs, 1037.09/qtl of sugar) followed by the public sector sugar factories (Rs. 1020.27/qtl. of sugar) and co-operative sector sugar factory (Rs. 956.78/ qtl of sugar). Among the below average factories, it was the highest in public sector (Rs. 1080.53/qtl. of sugar), followed by private sector factories (Rs. 930.83/qtl. of sugar production) and co-operative sector sugar factories (Rs. 010.67/qtl of sugar).

TABLE 5.15

CAPACITY UTILISATION OF SELECTED SUGAR FACTORIES FROM 1990-91- TO 2004-5

YEARS	*CO-OPERATIVE SECTOR*								
	ABOVE AVERAGE				*BELOW AVERAGE*				*AVG*
	HIRANYA-KESHI	*MALA-PRABHA*	*DOOD-GHANGA*	*AVG*	*PANDAVA-PURA*	*GHATA-PRABHA*	*HEMA-VATHI*	*AVG*	
1990 - 91	4982.56	3518.88	2004.46	3501.97	1057.00	1295.12	1135.09	1162.40	2332.19
1991 - 92	4620.95	3375.01	1859.70	3285.22	1023.00	1301.01	1189.03	1171.02	2228.12
1992 - 93	4829.17	3195.50	2098.12	3374.26	1025.00	1409.94	1133.59	1189.51	2281.89
1993 - 94	4872.33	3403.37	2112.27	3462.66	1008.00	1523.37	863.60	1131.66	2297.16
1994 - 95	4753.95	3629.10	2206.54	3529.86	1091.00	1580.86	1130.72	1267.53	2398.70
1995 - 96	4576.99	3220.21	2401.25	3399.48	992.00	1608.07	1000.79	1200.29	2299.88
1996 - 97	4525.95	3031.38	2198.24	3251.86	942.00	1376.75	919.56	1079.44	2165.65
1997 - 98	4946.99	3434.10	3211.02	3864.04	0.00	1496.87	843.59	780.15	2322.10
1998 - 99	5029.42	3604.83	3299.77	3978.01	1002.00	1097.19	1080.91	1060.03	2519.02
1999 - 00	4795.34	3511.98	3574.67	3960.66	1548.00	1568.43	110.23	1075.56	2518.11
2000 - 01	4703.03	3634.35	3688.62	4008.67	1588.00	1636.63	114.27	1112.97	2560.82
2001 - 02	4990.74	3396.19	4055.73	4147.55	1097.00	1998.34	819.77	1305.04	2726.30
2002 - 03	5107.39	3713.83	4101.32	4307.51	1213.00	1816.26	1068.96	1366.07	2836.79
2003 - 04	4426.29	3446.30	3626.70	3833.10	0.00	3253.14	1224.87	1492.67	2662.88
2004 - 05	5920.44	2877.60	3863.19	4220.41	0.00	1463.43	783.25	748.89	2484.65
AVG	4872.10	3399.51	2953.44	3741.68	905.73	1628.36	894.55	1142.88	2442.28

***Source:** Compiled and Analysed from the annual reports and general ledgers of above sugar factories from 1990-91 to 2004-05.*

Table 5.15a

CAPACITY UTILISATION OF SELECTED SUGAR FACTORIES FROM 1990-91- TO 2004-5

YEARS	PUBLIC SECTOR			PRIVATE SECTOR			OVERALL AVG
	ABOVE AVG	BELOW AVG	AVG	ABOVE AVG	BELOW AVG	AVG	(Co-operative, Public and Private)
	MY SUGAR	MYSORE		UGAR	DHAVANAGERE		
1990 - 91	2222.97	2337.01	2279.99	4088.88	1276.05	2682.47	2431.55
1991 - 92	2469.62	2379.34	2424.48	4112.77	1385.40	2749.08	2467.23
1992 - 93	2484.16	1802.23	2143.19	4550.01	1217.18	2883.60	2436.23
1993 - 94	2119.11	2025.17	2072.14	4148.00	1059.06	2603.53	2324.27
1994 - 95	3063.23	2208.83	2636.03	4592.06	1197.94	2895.00	2643.24
1995 - 96	2928.74	2147.71	2538.23	4496.09	1088.98	2792.54	2543.55
1996 - 97	1914.14	2934.29	2424.21	4934.06	1465.22	3199.64	2596.50
1997 - 98	2050.68	2263.25	2156.96	5595.82	1147.89	3371.85	2616.97
1998 - 99	3013.04	2202.66	2607.85	5241.96	1306.34	3274.15	2800.34
1999 - 00	3713.70	2279.21	2996.45	5523.34	1103.48	3313.41	2942.66
2000 - 01	3475.59	2242.23	2858.91	5859.33	2111.97	3985.65	3135.12
2001 - 02	3711.44	2526.79	3119.11	6945.54	2036.29	4490.92	3445.44
2002 - 03	3570.56	2415.94	2993.25	7527.16	2389.18	4958.17	3596.07
2003 - 04	2584.11	2329.73	2456.92	7077.55	2573.62	4825.59	3315.13
2004 - 05	2897.21	2364.95	2631.08	7978.40	1670.81	4824.61	3313.45
AVG	2814.55	2297.29	2555.92	5511.40	1535.29	3523.35	2840.52

Source: *Compiled and Analysed from the annual reports and general ledgers of above sugar factories from 1990-91 to 2004-05*

TABLE 5.16

CAPACITY UTILISATION OF THE SELECTED SUGAR FACTORIES DURING 2004-05

SECTORS	*FACTORIES*	*SUGARCANE CRUSHED (MT)*	*FOR 210 DAYS CAPACITY (T/210 DAYS)*	*DAILY CAPACITY (T/DAY)*	*CAPACITY UTILISATION*
	HIRANYAKESHI	509158.00	1050000.00	5000.00	48.49
ABOVE AVERAGE	MALAPRABHA	483436.00	735000.00	3500.00	65.77
CO-OPERATIVE S.F	DOODGHANGA	494488.85	735000.00	3500.00	67.28
	AVG	495694.28	840000.00	4000.00	60.51
	PANDAVAPURA	148012.00	735000.00	3500.00	20.14
BELOW AVERAGE	GHATAPRABHA	456371.00	462000.00	2200.00	98.78
CO-OPERATIVE S.F	HEMAVATHI	78325.00	262500.00	1250.00	29.84
	AVG	227569.33	486500.00	2316.67	49.59
OVER ALL CO-OPERATIVE S.F		361631.81	663250.00	3158.33	55.05
ABOVE AVERAGE PUBLIC SECTOR S.F	MY SUGAR	706920.00	1050000.00	5000.00	67.33
BELOW AVERAGE PUBLIC SECTOR S.F	MYSORE	503734.00	525000.00	2500.00	95.95
OVER ALL PUBLIC SECTOR S.F		554530.50	656250.00	3125.00	88.79

TABLE 5.16 contd...

SECTORS	*FACTORIES*	*SUGARCANE CRUSHED (MT)*	*FOR 210 DAYS CAPACITY (T/210 DAYS)*	*DAILY CAPACITY (T/DAY)*	*CAPACITY UTILISATION*
BELOW AVERAGE PRIVATE SECTOR S.F	UGAR	1635573.00	2100000.00	10000.00	77.88
ABOVE AVERAGE PRIVATE SECTOR S.F	DHAVANAGERE	157056.00	262500.00	1250.00	59.83
OVER ALL PRIVATE SECTOR S.F		526685.25	721875.00	3437.50	64.34
OVER ALL AVG		480949.19	680458.33	3240.28	69.40

Source: *Compiled and Analysed from the annual reports and general ledgers of above sugar factories from 1990-91 to 2004-05*

TABLE 5.17

COST STRUCTURE OF PROCESSING SUGARCANE INTO SUGAR IN SELECTED SUGAR FACTORIES

SECTOR	FACTORIES	CANE COST	OTHER COST	TOTAL	FIXED COST	GROSS COST	INCOME FROM BY PRODUCTS	COST OF PROCESSING
	HIRANYAKESHI	1071.38 (73.97)*	310.02 (21.40)	1381.40 (95.38)	66.98 (4.62)	1448.38 (100.00)	92.89	1355.49
ABOVE AVERAGE	MALAPRABHA	853.58 (46.89)	583.48 (32.05)	1437.06 (78.94)	383.48 (21.06)	1820.54 (100.00)	138.68	1681.86
CO-OPERATIVE S.F	DOODGHANGA	1083.74 (64.14)	276.9 (16.39)	1360.64 (80.53)	329.02 (19.47)	1689.66 (100.00)	342.91	1346.75
	AVERAGE	1002.90 (60.68)	390.13 (23.60)	1393.03 (84.28)	259.83 (15.72)	1652.86 (100.00)	191.49	1461.37
	PANDAVAPURA	800 (33.79)	16.8 (0.71)	816.8 (34.50)	1550.47 (65.50)	2367.27 (100.00)	195.00	2172.27
BELOW AVERAGE	GHATAPRABHA	1032.00 (89.62)	82.13 (7.13)	1114.13 (96.76)	37.36 (3.24)	1151.49 (100.00)	699.7	451.79
CO-OPERATIVE S.F	HEMAVATHI	900 (67.61)	321.101 (24.12)	1221.101 (91.73)	110.14 (8.27)	1331.241 (100.00)	165.60	1165.64
	AVERAGE	910.67 (56.33)	140.01 (8.66)	691.75 (42.79)	937.37 (57.98)	1616.67 (100.77)	353.43	1263.23
OVER ALL CO-OPERATIVE S.F		956.78 (58.53)	265.07 (16.21)	1042.39 (63.76)	598.60 (36.62)	1634.76 (100.38)	272.46	1362.30
ABOVE AVERAGE PUBLIC SECTOR S.F	MY SUGAR	960.00 (41.80)	1123.18 (48.90)	2083.18 (90.70)	213.58 (9.30)	2296.76 (100.00)	693.87	1602.89

TABLE 5.17 *contd....*

SECTOR	*FACTORIES*	*CANE COST*	*OTHER COST*	*TOTAI*	*FIXED COST*	*GROSS COST*	*INCOME FROM BY PRODUCTS*	*COST OF PRO-CESSING*
BELOW AVERAGE PUBLIC SECTOR S.F	MYSORE	1080.53 (54.36)	563.83 (28.37)	1644.36 (82.73)	343.38 (17.27)	1987.74 (100.00)	343.83	1643.91
OVER ALL PUBLIC SECTOR S.F		1020.27 (48.08)	843.51 (38.63)	1863.77 (86.71)	278.48 (13.29)	2142.25 (100.00)	518.85	1623.40
ABOVE AVERAGE PRIVATE SECTOR S.F	UGAR	1143.35 (61.79)	353.24 (19.09)	1496.59 (80.88)	353.89 (19.12)	1850.48 (100.00)	243.06	1607.42
BELOW AVERAGE PRIVATE SECTOR S F	DHAVANAGERE	930.83 (36.28)	351.93 (13.72)	1282.76 (50.00)	1282.76 (50.00)	2565.52 (100.00)	1574.19	991.33
OVER ALL PRIVATE S.F		1037.09 (49.03)	352.59 (16.40)	1389.68 (65.44)	818.33 (34.56	2208.00 (100.00)	908.63	1299.38
OVER ALL		1004.71 (51.88)	487.05 (23.75)	1431.95 (71.97)	565.13) (28.16)	1995.00 (100.00)	566.65	1428.36

Source: *Compiled and Analysed from the annual reports and general ledgers of above sugar factories from 1990-91 to 2004-05*
** figures in the parentheses denote per cent's to the total*

Among the above average factories cost of cane for the production of every quintal of sugar was the highest in private sector factories (Rs. 1143.35) followed by the co-operatives (Rs. 1002.90 /qtl of sugar) and public sector (Rs. 960/qtl of sugar). With respect to the expenses of other variable costs, it was the highest in public sector sugar factories; (Rs. 843.51) followed by the private sector sugar factories (Rs. 352.59) and co-operative sector sugar factories (Rs. 265.07/qtl). In the case of above average sugar factories, variable costs is the highest (Rs. 123.18/ qtl) followed by co-operative sector (Rs. 390.13/ qtl) and private sector (Rs. 353.24/qtl). It was observed that other variable costs is the highest in Mysugar factory (Rs. 1123.18/ qtl, above average public sector) followed Malaprabha sugar factory (Rs. 583.48 /Qtl) and Mysore Paper Mills sugar factory (Rs. 563/qtl).

CHAPTER 6

COMPARATIVE ANALYSES OF FINANCIAL PERFORMANCE OF SUGAR INDUSTRY IN KARNATAKA

The previous chapter discusses the results of the financial performance of sugar industry in Karnataka. The work on financial performance of sugar industry lends itself in Karnataka more of objectivity and further through probing with comparative analyses. The present chapter with comparative approach throws light on the causes responsible for the major trends observed in previous chapter. The comparative analysis of the major trends in terms of causes as well as their interaction among public sector, private sector and joint sector are made. It is hoped that, the comparative analysis among public, private and joint sector sugar industry would help in identifying some policy measures, which could be adopted to overcome the constraints from which the sugar factories suffered.

The comparative analyses among the public, private and joint sector sugar industry is organized under following headings:

6.1 Pattern of ownership and performance of sugar industry in Karnataka

6.2 Performance and growth of sugar factories

6.3 Financial performance of sugar factories

6.4 Statistical analyses of the sugar factories

6.5 Capacity utilisation of sugar factories

6.6 Cost of processing sugarcane into sugar

6.7 An overview of the functioning of the sugar factories SWOP (T) management technique'

6.1. Pattern of ownership and performance of sugar industry in Karnataka

In Karnataka out of the 52 sugar factories, 18 are under co-operative sector, 32 under are private sector and two under public sector. The main aim of the sugar factories under co-operative sector is self-help with minimum profit and maximum benefit to the members.

The data on sugar production, sugarcane crushed, average duration of crushing and recovery percentage of sugar from sugarcane was collected pertaining to the sugar factories as a whole in the state from 1990-91 to 2004-05. The data were categorised based on the ownership of the factories and analysed.

6.1.1 *Sugar production in Karnataka*

Since the number of sugar factories and the installed capacity to crush sugarcane in the existing sugar factories has increased over the study period, an overall growth rate of 4.34 per cent was observed in the total sugar production in the state. This might also be due to the increase in the availability of sugarcane over the years in the area.

The number of sugar factories, mainly in the private sector, has increased considerably in the state. This acted as impetus in increasing the sugar production in this sector from 37.84 lakh tones in 1990-91 to 64.78 lakh tones in 2004-05.

On the other hand the private sugar factories increased in number and their crushing capacities also increased, which resulted in increasing sugar production over the study period.

The public sector sugar factories showed a relatively steady contribution in sugar production among the three sectors. The installed capacity and the number of sugar factories under this sectors remained unchanged throughout the study period.

6.1.2 *Sugarcane crushed in Karnataka*

Similar to the sugar production in the state, the sugarcane crushed also showed an increasing trend over the study period. Here it can be recalled that the number of sugar factories were increased in that period.

A growth rate of 11.97 per cent annually in the co-operative sector could be attributed to the increase in the number of sugar factories in this sector. However, the co-operative sector sugar factories could increase their cane crushing capacity and utilisation of the installed crushing capacity, ultimately increasing the quantum of sugarcane crushed by them.

6.1.3 *Average duration of crushing*

The duration of crushing (days) mainly depends on the availability of sugarcane, the manpower and other raw materials as well as the installed capacity of the sugar factories to crush the sugarcane.

Due to non-availability of sufficient sugarcane to the factories, the factories in all the sectors could not crush for more than 199 days.

6.1.4 *Recovery percentage of sugar from sugarcane*

The recovery percentage of sugar from sugarcane heavily depends on the variety of sugarcane, efficient sugarcane crushing by the sugar factories and also the sucrose content in sugarcane at the time of crushing.

An average recovery of 10.46 per cent was found over the study period, which was optimum recovery percentage. A recovery per cent of more than 11 would reflect efficient procurement scheduling by the sugar factories and also the variety of cane produced in the hinterland of the sugar factory.

In line with these reasons, the recovery percentage was more than ten per cent in the co-operative, followed by the private sugar factories.

Since some co-operative factories were newly opened, using the latest technology of juice (sucrose) extraction, the co-operative sugar factories in aggregate could recover on an average 10.63 quintals of sugar from every 100 quintals of sugarcane. This was closely followed by Private sugar factories, where in procurement scheduling of sugarcane was efficient.

6.2 Performance and Growth of Sugar Factories

The basic information on ten selected sugar factories were colleted from 1990-91 to 2004-05 and were analysed pertaining to both physical and financial indicators. This information was further stratified into different strata based on the type of ownership. The physical and financial indicators selected were helpful reflecting the structure and performance of these sugar factories.

6.2.1 *Physical indicators*

6.2.1.1 *Installed capacity*

From Appendix-5 and 5a, it could be seen that, the private sector sugar factories had the highest installed capacity throughout the study period, followed by co-operative sugar factories. This was mainly because, the private sugar factories had more flexibility to increase their installed capacity based on their prediction of increase in sugarcane production and availability in the near future. They plan well in advance with regard to the variety and quantity of sugarcane availability and increase their capacity to crush the cane. In this regard, the private sugar factories had efficient managerial abilities, followed by the sugar factories under co-operative sector. This in turn was reflected in the installed capacity of overall sugar factories and the above average sugar factories in particular.

Among the co-operative sugar factories, the installed capacity was higher in the case of above average sugar factories. This was attributable to the availability of sufficient cane in the hinterlands of above average sugar factories. Similar trends were observed in the case of sugar factories under private and public sector sugar factories.

6.2.1.2 *Sugarcane Crushed*

The quantity of sugarcane crushed by a sugar factory depends on its installed capacity to crush cane, availability of required quantity of cane in required time and the availability of manpower and other raw materials. Due to higher installed capacity (Appendix-6 and 6a) and efficient management, the above average sugar factories could crush higher quantities of cane than their counterparts.

6.2.1.3 *Recovery Percentage of Sugar*

The recovery percentage of sugar from sugarcane depends on the sucrose content in sugarcane at the time of crushing, variety of sugarcane and the efficiency of factory to extract sugar from sugarcane.

A recovery of ten per cent sugar from sugarcane was found to be satisfactory, as opined by the authorities of different sugar factories. This level of recovery percentage was attained by all the sugar factories under study except below average co-operative sugar factories and Mandya sugar factory (above average public) for the study period of revealed by Appendix-7 and 7a.

Due to availability of high sucrose content varieties of sugarcane and efficient procurement scheduling, the recovery of sugar from sugarcane was the highest in the case of Ugarkhurd sugar factory (above average private).

However, on an average, the sugar recovery percentages were satisfactory throughout the study period.

6.2.1.4 *Duration of Crushing Season*

The duration of crushing season mainly depends on the installed capacity to crush sugarcane and the availability of sugarcane in time and in required quantity.

Due to higher installed capacities and availability of sufficient and suitable varieties of sugarcane, the above average sugar factories could prolong the crushing season above six months. On the other hand, below average sugar factories could not prolong the crushing season, due to lack of sufficient quantity of sugarcane with high sucrose content at different dates. In these cases, procurement scheduling was very difficult, as different varieties of cane were not available.

Planting different varieties of sugarcane and also staggering the planting in the hinterlands of each sugar factory could enable the sugar factories to prolong the duration of crushing, which in turn could help in higher levels of sugar production.

6.2.1.5 *Sugar Production*

Significant negative growth rates were observed (Appendix-9 & 9a) in all the sugar factories, except Hiranyakeshi (above average co-operative) and Ghataprabha (below average co-operative sector).

Because of the higher installed capacity (Appendix-5 & 5a) availability of sufficient sugarcane for crushing (Appendix-6 & 6a) and efficient management, the above average sugar factories could produce more sugar as compared to their counterparts.

Due to the same reasons, among the above average sugar factories, private sugar factory (Ugarkhurd Sugar Factory) had the highest sugar production followed by co-operative sugar factory.

6.2.2 *Financial indicators*

6.2.2.1 *Share Capital*

The share capital of an enterprise is important variable to be considered in the long run analyses.The overall share capital position (Appendix-10 & 10a) increased from Rs. 508.23 lakhs during 1990-91 to Rs. 1464.36 lakhs during 2004-05, with an annual growth rate of 7.29 per cent.The share capital increased at 15.05 per cent per annum in the case of private sector sugar factories followed by co-operative and the public sector sugar factories, which indicated the interest of the shareholders in the performance of the factories. In general, increase in the share capital of a business organisation is a welcoming feature.

6.2.2.2 *Reserve Funds*

Similar to the growth in share capital, the overall reserve funds (Appendix-11 & 11a) increased from Rs. 545.87 lakhs during 1990-91 to Rs.998.04 lakhs during 2004-05 with an annual growth rate of 6.56 per cent. The reserve funds increased at significant growth rates of 5.83 and 15.59 per cent per annum in the cases of co-operative and private sector sugar factories, respectively. But it was negative in case of public sector (-4.73), the accumulation of reserve funds was the highest in private sector sugar factories to meet the unforeseen expenses.

The accumulation of reserves was higher in the case of above average sugar factories as compared to their counterparts, as the above average sugar factories had more transaction and higher fixed assets (Appendix-14 & 14a). However, the build-up of reserves was satisfactory in general, except in the case of below average sugar factory.

6.2.2.3 *Owned Funds*

The owned funds of an organisation represent its financial strength as opined by the All India Rural Credit Review Committee Report.

The paid up share capital, reserves and other funds and net profit constitute the owned funds. The owned funds grew at an average annual growth rate of 6.95 per cent (Appendix-12 & 12a), due to significant growth rates of paid up share capital (Appendix-10 & 10a) and the reserve funds (Appendix-11 & 11a). The accumulation of owned funds over the study period was significant in all the three sectors. However, the growth rate was the highest in the case of private sugar factories, due to high growth rate of both paid up share capital and reserve funds of these factories. The co-operative sugar factories and the pubic sector sugar factories followed this. Growth of owned funds in Sugar factories were found among the below average sugar factories as well as the above average sugar factories.

6.2.2.4 *Assets position*

The current assets (Appendix-13 & 13a), fixed assets (Appendix-14 & 14a) and the total assets (Appendix-15 & 15a) had significant positive growth rates. The current assets and the total assets grew significantly in all the three sectors (based on ownership), whereas, the growth of fixed assets was significant only in co-operative and private sugar factories. The negative and non-significant growth rate in the case of public sector sugar factories was due to very slow growth rate in Mandya sugar factory and decrease in the fixed assets in the case of Davangere sugar factory. On the whole, the asset position was very strong in the case of co-operative sugar factories, followed by private and public sectors sugar factories, which indicated that the co-operative sugar factories on an average were found enough with respect to their assets position.

All the three types of assets under study *viz.*, current assets, fixed assets and total assets grew at significant rates. These three types of assets were more in above average sugar factories than their counter parts indicating higher investments in the assets in the case of above average sugar factories.

6.2.2.5 *Liabilities Position*

Similar to the assets position of sugar factories, the current liabilities (Appendix-16 & 16a), the borrowings (Appendix-17 & 17a) and the total liabilities (Appendix-18 & 18a) grew at significant rates in all the sugar factories under the present study. Even though the assets position was found to be very sound in the case of private sugar factories, these sugar factories had higher liabilities too, resulting in better management of assets and liabilities by external funds. A higher current liability, reflected higher outstanding expenses and payments to be made, which was the highest in co-operative sugar factories, whose current assets were also high, resulting in better management of sundry payable and receivables. An overall look into the assets and liabilities position revealed that, higher liabilities, coupled with sound assets position enabled better management of finances, increasing the volume of transactions.

6.2.2.6 *Total Sales*

The total sales increased from Rs. 3100.77 lakhs during 1990-91 to Rs. 7677.99 lakhs during 2004-05 (Appendix-19 & 19a), with significant growth rate of 6.51 per cent per annum. In general, in line with the production of sugar and sugar inventory maintained by the sugar factories, than their counter parts.

The values of total sales increased at a faster rate in above average sugar factories in general and Ugarkhurd (private) sugar factory in particular. This was mainly because of higher levels of sugar production.

In general, the total sales grew significantly in all the sugar factories over the study period.

6.2.2.7 *Gross Income and Total Expenses*

Since the volume of business was higher in the case of above average sugar factories, the gross income (Appendix-

20 and 20a) and total expenses (Appendix-21 and 21a) were also higher as compared to their counterparts. However, the growth rate of gross income and total expenses were significant in all the sugar factories considered in the present study.

In line with the assets and liabilities position and also the volume of sales, the private sugar factories had higher gross income and total expenses as compared to their counterparts and the other sectors.

The overall growth rate of 6.97 per cent per annum in gross income and 6.44 per cent per annum in total expenses indicated high volume of business carried out. This is a welcoming feature for a business organisation.

6.2.2.8 *Net Profit*

From Appendix-22 and 22a, it could be seen that overall net profit was negative (net loss) for six years and positive (net profit) for other nine years.

6.3 Financial Performance of Sugar Factories

The analyses of financial statements employing ratio analyses technique and their comparison indicate the extent of success of the organisation in its business performance. Different dimensions of financial performance like liquidity, solvency, turnover, efficiency and profitability along with the tests of financial strength of the sugar factories under the present study were analysed by employing different ratios and are discussed below under different sub-heads.

6.3.1 *Test of liquidity*

The liquidity ratios measure the ability of an organisation to cover its short-term obligations out of its short-term resources. The results are presented in Table 5.1.

6.3.1.1 *Current Ratio*

The current ratio was widely used for all the typical balance sheet ratios. This ratio was greater than unity in all the sugar factories. However the overall average ratio was 2.08. It may be pointed out that; a ratio of 2:1 is preferred in the case of joint stock companies. However, it was observed that the ratio above unity indicates a firm's ability to meet current obligations. This ratio was found to be greater than unity in all the sugar factories in the present study. Thus, it could be concluded that all the sugar factories had maintained a reasonable level of liquidity position.

6.3.1.2 *Acid Test Ratio*

This ratio is a stringent measure of liquidity, as inventories are excluded from current assets. A standard ratio of 1:1 is desirable for production units like sugar factories. The ratio was more than unity in all the sugar factories. This indicated that the sugar factories had maintained adequate cash balances to meet the day-to-day requirements.

6.3.1.3 *Liquid Assets to Total Assets Ratio*

This ratio was found to be 1.76, indicating that around 176 per cent of the total assets were made up of liquid assets.

This ratio was found to be the highest in private sector sugar factories (1.89), followed by co-operative sector sugar factories (1.72) and the public (1.67). A higher proportion of liquid assets in the total assets are a healthy sign, as any processing and trading organisation should have more working capital to meet the increased volume of crushing over the years.

6.3.1.4 *Inventory Ratio*

Inventory ratio measures the extent to which the net working capital finances the current assets. This ratio was more than unity in all the sugar factories, the overall weighted average being 2.08 per cent. A ratio of greater than one indicated that the net working capital was tied up in inventory.

This ratio was the lowest, even though greater than unity, in private sector sugar factory, followed by co-operative and public sector sugar factories, indicating that the co-operative sugar factories depended more on the net working capital to form the current assets.

6.3.2 *Tests of Solvency*

Solvency is the reflection of financial strength of an organisation to meet its medium and long-term obligation; the margin of safety offered to the creditors and the potential earnings from the use of borrowed funds. The solvency of the sugar factories was measured in terms of three types of solvency ratios and the results are presented in Table 5.2.

6.3.2.1 *Total Liabilities to Owned Funds Ratio*

The ratio of total liabilities to owned funds indicates the extent of debts (in rupees) per rupee of owned funds. This ratio varies between 1.07 (Hemavathi S.F) and 29.66 (Hiranyakeshi S.F), which revealed the reliance of these sugar factories more on external funds.

A value over unity of this ratio for a non-banking institution indicates a poor financial structure. However, the prescribed norm is 3:1. This ratio was not within the prescribed norm in most of the sugar factories. The overall weighted average ratio was also within the norm. Increased external debts, accumulated loss, debt servicing and lack of managerial ability to create owned funds were the root causes for the problem.

6.3.2.2 *Fixed Assets to Owned Funds Ratio*

The ratio of fixed assets to owned funds explains the extent of owned funds invested in fixed assets. It indicates that higher ratio is associated with problems of liquidation, because, the claims of the owners have to be met by sale of fixed assets, which are in non-liquid form

In the present study, this ratio was found to be more than unity in all the individual sugar factories except Ghataprabha (below average co-operative) sugar factory. However, the overall weighted average was found to be more than unity (4.35).

6.3.2.3 *Debt Equity Ratio*

The debt equity ratio explains the extent of external dependence of the organisation in relation to its equity. A ratio of less than 3:1 signifies the financial soundness of the organisation regarding its debts and equity positions. In the present study the debt equity ratio was far below the norm of 3:1 in Ghataprabha (1.15) and Mysore paper mills (1.06). This ratio indicated financial instability and unsoundness in all the sugar factories.

6.3.3 *Tests of Turnover*

The turnover ratio (activity ratio) reflects how efficiently the sugar factories maintained their resources in the key areas of sales, working capital and the inventory management. The ratios are presented in Table 5.3.

6.3.3.1 *Inventory Turnover Ratio*

The turnover ratio is an important tool to signal the positive and negative aspects of the economic health of an enterprise. This ratio was employed to measure the effectiveness of the sales efforts of the sugar factories.This ratio was more than 2.5 in Malaprabha (23.4, above average co-operative) and Mysugar (2.54, above average public sector), indicating high stock exit rate and better inventory management, as compared to other sugar factories. Due to high exit rate of the stock and its management, this ratio was higher in above average co-operative sugar factories than their counterparts.

6.3.3.2 *Total Assts Turnover Ratio*

Total assets turnover ratio, which is a ratio of annual sales to the total assets, indicates the sales efforts of the sugar factories, in relation to the total assets. Generally speaking, the higher the turnover, greater the efficiency and higher the rates of profit.

In the present study, this ratio was higher in public sector sugar factories, followed by private sector and co-operative sector sugar factories, indicating higher efficiency of turnover in the case of public sector sugar factories, as compared to private sector and co-operative sugar factories.

6.3.3.3 *Working Capital Turnover Ratio*

This ratio measures the efficiency with which the working capital was employed in the sugar factories. A greater working capital turnover ratio implies greater efficiency in utilising the working capital of an organisation in general. In the present study, the overall weighted average ratio of 4.28 was observed, which is satisfactory. This ratio was greater than 1.5 in all the sector of sugar factories, indicating greater efficiency of these sugar factories in utilising the working capital.

6.3.4 *Tests of Efficiency and Profitability*

The degree of efficient utilisation of resources and the profitability levels of sugar factories were measured by employing different ratios. The results are presented in Table 5.4.

6.3.4.1 *Efficiency of Capital Ratio*

A higher efficiency of capital ratio indicated efficient utilisation of capital. In the present study, public sector sugar factories had higher ratio as compared to co-operatives and private sugar factories. However, this ratio was less than 0.5 in all the sugar factories except Mysugar, Mandya (0.64, above average public sector).

6.3.4.2 *Gross Ratio (percentage)*

This ratio measured the total expenses for every Rs.100 gross income. The overall weighted average ratio was 292.87 per cent, indicated that the factory on an average incurred total expenses of Rs.292.87 for every Rs.100 gross income. A ratio more than 100 indicated that the total expenses were more than the gross income. This ratio was the highest in

Pandavapura (Rs. 3118.77, below average co-operative) followed by Hemavathi co-operative sugar factory (251.59). This ratio was the lowest in private sugar factory, indicating higher net profit.

6.3.4.3 *Net Profit to Owned Funds Ratio*

The net profit to owned funds ratio was –6.44, which indicated that, the rate of returns on owned funds were negative. Since the co-operative sugar factories were under loss over the study period, this ratio was found to be negative. On the other hand, because of high net profit in relation to the growth in the owned funds, the private sugar factories could manage to get the returns to the tune of 108.4 per cent of the owned funds. This ratio was the highest in Ugarkhurd (above average private) sugar factory, indicating its efficiency to get higher returns on owned funds. Though this ratio is a reliable measure of the performance of a business organisation, it is considered only as a supplementary measure in case of service oriented institutions (co-operative and public sector sugar factories), because of their stress on social objective.

6.3.4.4 *Net Profit to Total Assets Ratio*

The net profit to total assets ratios were negative in public sector sugar factories and the highest ratio was registered by co-operative sector sugar factories.

In all the sugar factories, this ratio was very low, which indicated marginal profit. The profit level was low in relation to total assets. Hence, efforts should be made to increase profit by decreasing expenditure as sales have increased considerably over years in all the sugar factories. Return on asset is one of the best standards of evaluation.

Net profit to total assets should be more than two per cent for an efficient utilisation of total assets. Private and co-operative sectors sugar factories were more efficient in utilising the total assets.

6.3.4.5 *Net Profit to Total Sales Ratio*

The ratio of net profit to sales indicates the margin per unit of sugar sales. In the present study, the ratio gave a dismal picture, particularly in co-operative and public sector sugar factories. This indicated that the earning capacity of factories was very unsatisfactory. This low margin of profit earnings was also due to Government's dual price policy and monthly release mechanisms for sugar in the country.

6.3.5 ***Tests of Financial Strength***

The long run liquidity position of any organisation would be reflected by the net capital ratio. It would be considered as a long-term concept, along with net worth and equity capital ratio, as it involved the value of real estate. The results of these tests are presented in Table 4.5.

6.3.5.1 *Net Worth*

The average net worth of the sugar factories over the study period was found to be Rs. 7558.97 lakhs. Due to creation of huge assets with less borrowing, the co-operatives had better net worth than public and private sector sugar factories. The net worth was negative in the cases of Mysugar, Mandya (above average public) sugar factory, indicating that, the total assets were not sufficient to pay off all the liabilities of this factory. On the other hand among the above average sugar factories, the co-operative sugar factories had the highest net worth, followed by the private sector sugar factory, whereas, among the below average sugar factories, the highest net worth was found in the public sector sugar factory, followed by the co-operative sugar factories.

6.3.5.2 *Net Capital Ratio*

From Table 5.5, it could be seen that the net capital ratio was more than unity in all the sugar factories except Hemavathi and Mysugar sugar factories. This indicated that, in these two sugar factories, the assets were not sufficient to cover all the liabilities. Due to sound assets position, the public sector sugar factories had higher net capital ratios than co-operative and private sector sugar factories.

6.3.5.3 *Equity Capital Ratio*

An equity capital ratio of more than 3:1 was achieved by private and public sector sugar factories, whereas it was 0.93 in the case of co-operative sector sugar factory, indicating lower build-up of equity in the capital. However, an overall average ratio of 0.98 was found, which was satisfactory.

6.3.6 *Fixed Assets Ratio*

The fixed assets ratios *viz.*, fixed assets to total assets ratio, fixed assets to net worth ratio and fixed assets to total sales ratio were computed and are presented in Table 5.6.

From the table it could be seen that, the public sector sugar factories had higher ratios as compared to co-operative and private sugar factories, indicating the creation of sufficient fixed assets, as proportion of total assets, net worth and the total sales in public sector sugar factories.

The overall average fixed assets to total assets ratio was 1.53, which indicated that fixed assets formed 153 per cent of the total assets. This ratio was 113 per cent in the case of co-operative sugar factories followed by private sector sugar factories. Similar trends were found in the case of fixed assets to net worth and fixed assets to total sales ratios.

6.3.7 *Dupont Analyses*

The Dupont Formula allows you to examine how a firm generates its return on equity. It links several critical ratios, If operating management is strong (high PM), and asset management is strong (high TAT), and capital structure management is strong (appropriately low EM), then ROE will be high for sound business reasons and the sugar industries said to be in strong financial condition. ROE can be high for the wrong reasons however. For example, when ROE is high because the equity multiplier is high, this means that high returns are really coming from overuse of debt, which can spell trouble for the sugar factories. Return on equity of public sector sugar factories has been negative till the year 2003-04 from the year 2000-01. Public sector did well in the beginning that is in the year 1991-92 and after that earned negative return on equity for 2 years and increased its return on equity for continuous three years and ROE falls continuously till the year 2003-04 this is due to weakness of operating management and asset management

6.3.8 *Break Even Analyses*

Break-Even analyses is useful in understanding the relationship between level of operation, revenue generation, variable cost and fixed costs and their influence on profitability. As such, it has a wide variety of uses in decision-making. Although Below average co-operative sector of sugar factories has achieved BEP sales in which they are all incurring no loss no profit. It shows that there was fluctuation of income to the sugar factories over the study period.

6.3.9 *Safety Margin Analyses*

Safety Margin is a management control technique which emphasizes to decrease or increase the production of sugar to achieve breakeven sales at which sugar factory earns no profit/ no loss. Public Sector is more predominant in achieving Safety Margin, which is expressed in percentage of breakeven sales. Safety Margin is positive in the beginning but the latter the years Safety Margin is decreased.

6.4 Statistical Analyses of the Sugar Factories:

The important statistical used here are:

6.4.1 *Standard Error*

The standard error of a statistic is the standard deviation of the sampling distribution of that statistic. Standard errors are important because they reflect how much sampling fluctuation a statistic will show. Standard error of the borrowings is in increasing trend in the sugar industry of Karnataka. Standard error of the total expenses is in increasing trend in the sugar industry of Karnataka and in the last 4 years it shows declining trend. Standard error of the current liabilities increased from year to year in 15 years of time because of borrowing is increased from year to year. Standard error of the total asset position is more in the beginning and further is reduced over 14 years of time.

6.4.2 *Estimation of Standard Error*

The main objective of much statistical investigation is to make predictions preferably on the basis of mathematical equation. Here least square method is used in which the dependent variables is to be predicted in terms of a single independent variable. In many problems of this kind the independent variable is observed without error or with the error in the dependent variable. In the relationship between X and the mean Y is linear or close enough to a straight-line so that least square line yields reasonably good predictions. estimated standard error of the sugar production is more in the My Sugar factory of Public sector compared to other co-operative and private sectors and estimated standard error of sugar production is low compared to co-operative sugar factory having the estimated value ranging from 25 per cent to 50 per cent. Estimated standard error of the recovery of sugar is more in Pandavapura sugar factory of below average co-operative sector compared to any other sugar factory of sugar industry in Karnataka. This is due to various factors like company not adopting diffuser mechanism, which increases the yield of sugar recovery. Estimated Standard error of the Borrowings is more in case of above average and below average co-operative sector sugar factories followed by public sector and private sector sugar factories.

6.4.3 *t-Test*

t-test is a test of hypothesis use to determine the significant difference between sample mean and population mean by holding tentative statements which are to be tested by the researcher with the degree of freedom. As t-value calculated and computed in the table for various selected physical and financial indicators of the project are considered to be significant difference in their mean values if t- statistic value is $|t|<$ table value (2.145) at 5 per cent level of significance. If computed t-values are less than table value then the difference is not significant other wise it is significant. Level of reserve funds and total liabilities of public sector have shown significant difference from its mean value as table value is more than calculated t-value i.e. Appendix-11 (Level of Reserve Funds) and 18 (Total Liabilities).

6.4.4 *Compound Growth Rate*

Growth rates are calculated as annual averages and represented as percentages. Except where noted, growth rates of values are computed from constant time series. Three principal methods are used to calculate growth rates: least squares, exponential endpoint, and geometric endpoint. Rates of change from one period to the next are calculated as proportional changes from the earlier period. Compound growth of sugarcane crushed incase of co-operative sector sugar factories is less than zero although the installed capacity compound growth rate is about 2.5 per cent. This is because numbers of crushing days are less compared to public sector and private sector. In case of private sector sugarcane crushing is more and followed by public and private sector sugar factories. Compound growth rate of sugar production is more. Incase of private sector which accounted for just 2.5 per cent growth compared to other co-operative and public sector accounting for below zero per cent of growth in the sugar production. The very less sugar production is observed in case of public sector of sugar factories. CGR of level of current liabilities is more in case of public sector sugar factories accounting for 15 per cent growth and followed by private sector accounting for 6 per cent growth and co-operative sector resulted in 3 per cent growth. CGR of level of total sales was more in case of private sector accounting for 8 per cent growth and followed by 7 per cent growth of public sector and 6 per cent growth of co-operative sector sugar factories.

6.5 Capacity Utilisation of Sugar Factories

Capacity utilisation was one of the important indicators of operational efficiency of a processing plant. It is generally recognised that a high level of capacity utilisation was imperative for reducing the costs of production. The year wise capacity utilised by the sugar factories is indicated in Tables 5.37. 5.37a and 5.38. It can be observed from the table that on an average 69.40 per cent of the installed capacity was utilised.

Since the cane supply was not uniformly distributed throughout the crushing season, hence the sugar factories could not achieve the highest capacity utilisation of the installed capacity. Among the sample sugar factories the public sector sugar factories could achieve the highest utilisation of the installed capacity, as compared to co-operative and private sector sugar factories. Among the above average sugar factories capacity utilisation of the installed capacity was the highest in private sector (Ugar, 77.88 per cent), followed by public sector (67.33 per cent) and co-operative sector (60.51 per cent).

Among the below average sugar factories, however, the average capacity utilisation was to the tune of 95.95 and 59.83 per cent in public and private sugar factories respectively, whereas it was 49.59 per cent in co-operative sector sugar factory. The main reasons for this low level of capacity utilisation were non-availability of different varieties of sugarcane with different maturity periods and inefficient cane procurement scheduling. The capacities utilised by these sugar factories for the study period are given in Tables 5.15. 5.15a and 5.16. From this table, it could be seen that the overall capacity utilisation was to the extent of 2840.51 tones per day. This was mainly because of higher quantity of cane available for crushing throughout the crushing season.

The utilisation of installed capacity was the highest in private sugar factory (3523.35 tones per day), indicating its efficient functioning as compared to public sector (2555 tones per day) capacity utilisation and co-operative sector sugar factories (2442.28 tones per day). Since the utilisation of the installed capacity was the highest in Ghataprabha sugar factory (260 per cent during 2003-04, below average co-operative) followed by Davanagere sugar factory (206 per cent inn 2003-04, below average co-operative) and Mysore paper mills sugar factory (101 per cent, below average public sector).

On the other hand, the factories utilising hardly 50 per cent of their installed capacities should think of providing incentives to the farmers to get more cane for crushing, there by increasing the utilisation of the capacity could be increased by efficient cane procurement scheduling.

6.6 Cost of Processing Sugarcane into Sugar

Cost structure reefers to the various cost components that are present (i.e. both fixed and variable costs) in converting sugarcane into sugar. The cost structure would essentially be different for each sugar factory, because of difference in the methods employed in sugarcane processing, operations, mechanical setup and differences in the exogenous factors like government taxes, concessions, license charges, policies, etc. The indigenous factors like the installed capacity and its utilisation, the cane price fixed by the factory, other developmental activities taken by the individual factory, also have bearing on the differences in the cost of processing by the sugar factory.

From Table 5.17, it could be seen that on an average, the cost of processing was Rs.1428.36 per quintal of sugar. The processing cost was the highest in the case of public sector sugar factories (Rs. 1623.40/qtl) followed by co-operative sector (Rs.1362.30/qtl) and private sector sugar factories (Rs.1299.38/qtl). Since the below average sugar factories had lower capacity utilisation (except Ghataprabha and Mysugar sugar factory) and as they did not have control over the exogenous factors, the cost of processing was higher than their counterparts. Among the below average sugar factories, the cost of processing was the highest in the case of public sector sugar factory (Mysore paper mills), since this factory crushed very little cane as reflected by the highest share of variable cost (82.73 per cent) in the total cost. This further was responsible for lower share (17.27 per cent) of fixed costs in the total cost.

In general, the cane cost formed the prime share in the total cost (51.88 per cent), followed by the fixed costs (28.16 per cent) and other variable cost (23.75 per cent). The sugar factories should constantly upgrade their technology to offset the exogenous factors. To faster this government should suitably modify its sugar policy so that the sugar factories can realize and retain a high profit to enable fresh investment. However, in the long run, the continuous increase in the cane prices should be discontinued to ensure the viability of sugar production. To help the cane growers realize higher incentives, better infrastructural facilities should be created for higher productivity in terms of cane and recovery percentage be realised through research and extension.

6.7 Overview of the Functioning of the Sugar Factories: SWOP (T) Management Technique

The SWOP (T) management technique was employed to examine the functioning of the sugar factories from four different broad angles, namely, strength, Weakness, Opportunities and Problems (Threats). Since all the above average sugar factories were characterised by somewhat similar strengths and weakness, and so also all the below average sugar factories, the analyses was done separately for above average and below average sugar factories, irrespective of their type of ownership. The results of the analyses are presented and discussed under appropriate heads.

6.7.1 *Above Average Sugar Factories*

6.7.1.1 *Strengths*

a. These sugar factories have higher level of installed capacity i.e., these sugar factories have installed the machinery to crush about 5000 tonnes of cane per day (TCD). Since the installed capacity was more, the cane crushed and the sugar production were higher than the below average factories.

b. The sugar recovery percentage by these sugar factories was satisfactory.

c. Due to availability of sufficient cane with higher sucrose content at the time of crushing, these sugar factories could prolong the duration of crushing for more than six months.

d. The share capital and reserve funds were found to be increasing steadily throughout the study period.

e. These factories had sound assets positions as compared to their respective counterparts, indicating higher investments in the assets.

f. In line with the volume of business and sugar production, the above average sugar factories had higher net profit than their respective counterparts.

g. The current ratio was found to be greater than unity in all these sugar factories.

h. The liquid assets to total assets ratio was higher in private sector sugar factories as compared to their counterparts, which is a healthy sign, as any processing and trading organisation should have more working capital to meet the increased volume of crushing over years.

i. The total liabilities to owned funds average ratio was found to be 7.41. were as the ratio was beyond the prescribed norm of 3:1 in all the sectors of sugar factories.

j. Due to high exit rate of the stock and its efficient management, the inventory turnover ratios were found to be better in above average public and private sector sugar factories than their counterparts. This was found to be in line with total assets turnover and working capital turnover ratios.
k. Due to economic of large-scale production, the unit cost of sugar production was less in above average sugar factories than their counterparts.

6.7.1.2 *Weakness*

a. The borrowings are increasing at an alarming rate in these sugar factories.
b. The inventory ratios were greater than unity in all these sugar factories, which indicated that the net working capital was tied up in inventories.
c. The ratio of total liabilities to owned funds was found to be beyond the prescribed norm of 3:1.
d. The efficiency of capital ratios in above average sugar factories were found to be less than their counterparts, which indicated inefficient utilisation of capital.
e. The utilisation of the installed capacity was found to be less than 70 per cent, which is less.
f. All the above average sugar factories were found to be functioning below their breakeven levels of sugar production.

6.7.1.3 *Opportunities*

a. An overall look into the assets and liabilities position revealed that, higher liabilities coupled with sound assets position enable better management of finances, thereby increasing the volume of transactions.
b. Since these sugar factories have installed the machinery to crush about 5000 tones of cane per day (TCD), there is vast scope to crush more and more cane, thereby to improve the financial position and achieve break-even level of production, improve

capacity utilisation and reduce the unit cost of sugar production.

6.7.1.4 *Problems (Threats)*

a. The net profit to owned funds ratios were found to be very low, which indicated lower rates of returns on owned funds. This was in line with other two types of profitability ratios *viz.*, net profit to total assets and net profit to total sales ratios.

b. The net worth was negative in all sugar factories, which was reflecting below unity of net capital ratio, this indicated that the assets were not sufficient to cover all the liabilities of the sugar factories.

c. The high rates of borrowing by these factories may reach alarming levels, which may lead to liquidation.

d. The major portion of net working capital was made up of the inventories. This may lead to non-availability of sufficient working capital for day-to-day expenses.

6.7.2 ***Below average sugar factories***

6.7.2.1 *Strengths*

a. The sugar recovery percentage was found to be satisfactory.

b. The share capital was found to be increasing steadily throughout the study period.

c. The current ratios were greater than unity in all the sugar factories.

d. The debt-equity ratio was found to be within the prescribed norm of 3:1 in co-operative and private sugar factories, which signifies the financial soundness of these sugar factories regarding their debts and equity positions.

e. Efficiency of capital ratios was higher in these factories than their counterpart, indicating efficient utilisation of capital.

6.7.2.2 *Weakness*

a. Low level of installed capacity.
b. Since the installed capacity itself was less, the cane crushed and the sugar produced by these factories were less than the above average sugar factories.
c. Due to non-availability of sufficient cane throughout the season, the factories could manage to crush only for five months in a year.
d. These sugar factories had lower asset levels as compared to their respective counterparts.
e. The borrowings were found to be increasing at significant rates in these sugar factories.
f. In line with the volume of business and sugar production, these sugar factories had lower net profit than their respective counterparts.
g. The utilisation of installed capacity was found to be less, which is unsatisfactory.
h. The unit cost of sugar production was higher in these sugar factories as compared to their counterparts.

6.7.2.3 *Opportunities*

a. The share capital was found to be increasing steadily. At times, the sugar factories can invite for more share capital.
b. There is scope to increase the cane crushing activities by these factories, since the capacity utilisation was found to be less and the factories were functioning below the break-even level of production. This intern improves the financial position of these factories and reduces the unit cost of sugar production.

6.7.2.4 *Problems (Threats)*

a. The acid test ratios were found to be more than unity in all these sugar factories, which indicated that these sugar factories had to maintain adequate cash balances to meet the day-to-day cash transactions;

instead of depending on their inventory for immediate financial commitments.

b. The ratios of total liabilities to owned funds was found to be beyond the prescribed norm of 3:1 in private and public sector below average sugar factories, indicating poor financial structure with regard to their liabilities and owned funds.

c. The net profit to owned funds ratios were found to be very low, which indicated lower rates of returns on owned funds. This was in line with other two types of profitability ratios *viz*. net profit to total assets and net profit to total sales ratios.

CHAPTER 7

POLICY RECOMMENDATIONS AND SUGGESTIONS

On the basis of the findings of the present study and Researcher's own observations during investigation on the problem, the following policy implications are drawn and appropriate strategies recommended for improving the performance of the sugar factories:

1. The minimum radial distance between an existing sugar factory and new factory may be 25 km, in place of the present 15 km. Additional requirement of sugarcane for expansion of capacity beyond 5000 TCD should accrue through increased productivity and not by expansion of the area under sugarcane.
2. The factories have to intensify work on research, development and extension activities to diffuse technical know-how of sugarcane production to the farmers, so as to enable the factories for efficient utilisation of installed capacity, increase duration of crushing over eight months, achieve break-even level of production, reduce unit cost of sugar production and improve profitability
3. State Government may consider the drought / flood affected factories may be allowed additional open market borrowings to help sugar factories to meet the fixed costs. The State government may offer this

as loan to sugar factories at a lower rate of interest per annum; the central government may provide interest subsidy to meet the difference between the coupon rates on the bonds raised through additional market borrowings.

5. For diversification of sugar industry in the state, following policy intervention by the Central Government are necessary for exploitation of by-products:
   - Oxygenation of gasoline may be made compulsory.
   - The Central Government may strictly implement its long-term policy for blending of petrol with ethanol.
   - Fiscal and other incentives provided (in regard to blending of petrol with ethanol) may be in place for a minimum period of five years, without any rollback.
   - Purchase price of ethanol may be fixed for a period of three years (with suitable price escalation clauses) in place of the present practice of fixing the price annually.
   - The Central Government may take steps to prevent frequent changes in PPAs for cogeneration projects.
6. National Energy Policy may have provisions for (i) mandating the proportion of 'green' power to be purchased in the overall power purchase in a state; and (ii) preferential tariff for cogenerated power. The Central Electricity Act may also be suitably amended to provide teeth to the policy to support 'green' power.
7. For ensuring higher cane availability to the factories, the sugar factories should supply the required inputs like seed sets, fertilizers and chemicals for the production of cane on credit basis and the cost of which may be recovered from the farmers at the time of procurement of cane from them.
8. In order to ensure that the Research & Development efforts get a proper direction and fillip in the State,

the efforts of the existing institutions may be adequately strengthened for increasing productivity in the fields as also in the mills by introduction of new technologies and also by effective extension mechanism. Sugar Development Fund may also assist project linked extension activities of these institutions.

9. Working capital to co-operative sugar factories may be made available through National Co-operative Development Corporation (NCDC) at a subsidised rate of interest. For this purpose, NCDC may arrange cheaper funds through External Commercial Borrowings (ECBs) and the Government of India may accord permission as well as provide guarantee in this regard.
10. Various experiments conducted under All India Coordinated Research Project (AICRP) have shown that the newly developed varieties are suitable to be grown under specific climatic conditions. Therefore only the recommended varieties are to be cultivated suitable to the regions. The farmers should be encouraged to cultivate improved varieties (like Co-740, Co-7219 and C0-62175) and high sucrose yielding varieties (like CoC-671 and CoC-92061) of sugarcane.
11. In order to improve profitability of sugar industry through value addition to byproducts like bagasse and molasses, co-operative sugar mills having potential for setting up of diversification projects may be encouraged. In this context, NCDC may also explore the possibility of arranging external commercial borrowings and Government of India may accord permission as well as guarantee in this regard.
12. Rate of interest on SDF loans may be so prescribed as to maintain a uniform concessional nature of these loans, that is, the rates of interest should increase or decrease in accordance with the interest rate regimes prevailing in the market.

13. The sugar production capacity created by these sugar factories by investing huge funds is under utilised. In order to realize the reasonable returns to the investments, the factories should make full use of the available capacity.
14. National Energy Policy may have provisions for (i) mandating the proportion of 'Green' power to be purchased in the overall power purchase in a state; and (ii) Preferential tariff for co-generated power. The Central Electricity Act may also be suitably amended to provide teeth to the policy to support 'green' power.
15. In order to ensure that the Research & Development efforts get a proper direction and fillip in the country, the efforts of the existing institutions may be adequately strengthened for increasing productivity in the fields as also in the mills by introduction of new technologies and also by effective extension mechanism. Sugar Development Fund may also assist project linked extension activities of these institutions.
16. Beyond the initial time limit, a maximum of 3 months may be permitted for lifting of levy sugar, where after, the levy sugar quota would automatically be converted into free sale sugar, without any recurring levy obligation on this portion of levy sugar.
17. Augmentation of sugar supply in the country may be effected when required through facilitation of imports of raw sugar. Indian Sugar EXIM Corporation (ISEC), an organ of the sugar industry, may play a prominent role both in import.
18. The availability of sufficient quantity of good quality sugarcane in the mill area is an important pre-requisite for the factory to be economically viable. Further, the cost of transportation and deterioration in quality increases significantly with the distance of the areas from where sugarcane is procured in the

state. It is therefore necessary that suitable steps be taken by the state government to ensure that sufficient sugarcane is developed and grown in the mill area for the purpose.

19. The incidence of woolly aphid as a new pest on sugarcane came to light in August 2002 in Belgaum district and moved swiftly to Bhadra canal areas and Cauvery basin in southern Karnataka. The incidence and alarming rate of spread and severity has created panic among the cane growers in Cauvery basin who have already suffered substantial losses due to drought during the previous years.
20. Lack of sufficient knowledge about the pest in the ecosystem and preparation to gear up to situation among the technologies has been a major concern to suggest effective management practices.
21. Sugar industry in Karnataka State need better infrastructure like good irrigation facilities, availability of power, properly maintained road for transportation of sugarcane from field to sugar mills etc. Inadequate infrastructure has adversely affected the yield and quality of sugarcane.
22. Sugarcane is a perishable crop, which must be crushed at the earliest after harvesting. For this, and other considerations, the sugarcane grower is obliged to supply cane to a particular mill. Therefore the sugar mill becomes a monophonic buyer. An enlightened mill owner may be expected to fix a remunerative price for sugarcane taking into account the relative prices of competing crops, as otherwise, the cane growers would divert the land to growing of other crops, thus adversely affecting the availability of sugarcane to the mill. However, in absence of any competing crops, the sugar mill may be tempted to fix an unduly low price, thus adversely affecting the earnings of the sugarcane growers.

23. The factories have to intensify work on research, development and extension activities to diffuse technical know-how of sugarcane production to the farmers, so as to enable the factories for efficient utilisation of installed capacity, increase duration of crushing over eight months, achieve break-even level of production, reduce unit cost of sugar production and improve profitability.
24. The sugar factories should maintain adequate cash balances to meet day-to-day requirements instead of depending on inventory for immediate commitments.
25. The evaluation studies on the working of these sugar factories may be entrusted periodically to autonomous research institutions, so that, the sugar factories can know their weaknesses and try to avoid them in order to increase the operational efficiency.

CHAPTER 8

CONCLUSION

Karnataka is the third largest producer of sugarcane in India. The number of sugar factories and the installed capacity to crush sugarcane in the existing sugar factories has increased over the study period. The basic information on ten selected sugar factories were colleted for 15 years from 1990-91 to 2004-05 and are analysed pertaining to both physical and financial indicators. The private sector sugar factories had the highest installed capacity throughout the study period, followed by co-operative and public sector sugar factories. Due to higher installed capacity and efficient management, the above average sugar factories could crush higher quantities of cane than their counterparts. The recovery percentage of sugar from sugarcane depends on the sucrose content in sugarcane at the time of crushing, variety of sugarcane and the efficiency of factory to extract sugar from sugarcane. A recovery of ten per cent sugar from sugarcane was found to be satisfactory, as opined by the authorities of different sugar factories but some of the sugar factories are not achieving this target. On an average, the sugar recovery percentages were satisfactory throughout the study period. The higher installed capacities and availability of sufficient and suitable varieties of sugarcane, the above average sugar factories could prolong the crushing season above six months. On the other hand, below average sugar factories could not prolong the crushing season, due to lack of sufficient quantity of sugarcane with high sucrose content at different dates. The higher installed capacity and availability of sufficient sugarcane for crushing and efficient management, the above average sugar factories could produce more sugar as compared to their counterparts.

The share capital increased at 15.05 per cent per annum in the case of private sector sugar factories followed by co-operative and the public sector sugar factories, which indicates the interest of the shareholders in the performance of the factories. The accumulation of reserves was higher in the case of above average sugar factories as compared to their counterparts, as the above average sugar factories had more transaction and higher fixed assets. However, the build-up of reserves was satisfactory in general, except in the case of below average sugar factory. The accumulation of owned funds over the study period was significant in all the three sectors. However, the growth rate was the highest in the case of private sugar factories, due to high growth rate of both paid up share capital and reserve funds of these factories. The co-operative sugar factories and the pubic sector sugar factories followed this. The current assets, fixed assets, and the total assets had significant positive growth rates. On the whole, the asset position was very strong in the case of co-operative sugar factories, followed by private and public sectors sugar factories, which indicated that the co-operative sugar factories on an average were found enough with respect to their assets position. All the three types of assets under study *viz.*, current assets, fixed assets and total assets grew at significant rates. These three types of assets were more in above average sugar factories than their counter parts indicating higher investments in the assets in the case of above average sugar factories. Similar to the assets position of sugar factories, the current liabilities, the borrowings and the total liabilities grew at significant rates in all the sugar factories. An overall look into the assets and liabilities position revealed that, higher liabilities, coupled with sound assets position enabled better management of finances, increasing the volume of transactions. In general, in line with the production of sugar and sugar inventory maintained by the sugar factories, the value of total sales were higher in the case of above average sugar factories, than their counter parts. The values of total sales increased at a faster rate in above average sugar factories. This was mainly because of higher levels of sugar production. In general, the total sales grew significantly in all the sugar factories over the study period. The growth rate of gross income and total expenses were significant in all the sugar factories as considered in the present study. The private sugar factories had higher gross income and total expenses as compared to their counterparts and the other sectors. It could be seen that overall net profit was negative (net loss) for six years and positive (net profit) for other nine years.

In an era where there is a need for inclusive growth, the sugar industry in Karnataka is amongst the few industries that have successfully contributed to the rural economy. It has done so by commercially utilising the rural resources to meet the large domestic demand for sugar and by generating surplus energy to meet the increasing energy needs of the state. In addition to this, the industry has become the mainstay of the alcohol industry. The industry is facing crisis due to over production of sugar for the last few years, which has accumulated into huge carry over stocks. Faced with a projected loss of Rs 480 crore for the current year, the sugar industry in Karnataka has urged both the state and central governments to grant few concessions. Unless the government helps the industry to over come the crisis, the industry will not be in a position to disburse the cane arrears. Cane arrears for year 2007 stands at Rs 480 crore. The prices are ruling low, while the cost of production is Rs 1,370 per quintal in Karnataka, the realisation is about Rs 1,070 forcing us to undergo a loss of nearly Rs 300 per quintal. Export of sugar is also not encouraging as the realisation is much lower at Rs. 1,000 per quintal. Therefore the state government to waive purchase tax on sugarcane and grant incentive of Rs 1,000 per tonne on exports as given by neighboring Maharashtra and levy zero per cent value added tax on sugar. There is considerable scope for further reduction in the cost of production of both sugarcane and sugar in India with liberalisation of controls on the sugar industry. Consolidation of land holdings and corporate farming on the raw material side and expansion of capacity on the unit size are important developments and would lead to substantial improvements in productivity, thereby rendering cost-effective producer of sugar in the world. There is considerable scope for further reduction in the cost of production of both sugarcane and sugar in India with liberalisation of controls on the sugar industry. Consolidation of land holdings and corporate farming on the raw material side and expansion of capacity on the unit size are important developments and would lead to substantial improvements in productivity, thereby rendering cost-effectiveness of production of sugar in the world.

Karnataka has 52 sugar factories both in co-operative, private and public sectors, to convert the irregularity in cash credit into medium term loan on soft terms, there may be rescheduling of the existing term loans into long-term loans at reduced interest rates. It is suggested that the government should declare sugar industry as priority sector as it would enable the banking sector to extend more credit to the industry. Like Brazil, out of the total cane available for crushing, 45 per cent goes for sugar production and 55 per cent for the production of ethanol directly from sugarcane juice. This gives the sugar industry an additional flexibility to adjust its sugar production keeping in view the sugar price in the international market. The area under sugarcane in the state is presently 1.8 per cent of total cultivable area in the state and about 3.2 per cent of the irrigated area. There is considerable scope for increasing the area under sugarcane considering the fact that it is more profitable compared to other crops. The sugar industry will not be lacking in meeting the requirement of ethanol. In a market economy, there would be a considerable shift from the *gur* and *khandasari* sectors, which are inefficient producers with poor quality. In the current scenario of glut in sugar production, it may be advisable to divert such additional cane for the production of alcohol after meeting the sweetener requirement. The task force on the sugar industry for the Tenth Five Year Plan has suggested the evolution of a national policy on alternative fuels, which would include the use of ethanol-blended gasoline. Until such a policy is evolved, sugar factories and distilleries should be encouraged to produce ethanol from the surplus alcohol available with them. Utilization of molasses for the production of ethanol will not only provide value-addition to the byproduct, it can also ensure better price stability and price realization of molasses for the sugar mills. This will improve the viability of the sugar mills, which will in turn benefit cane growers.

WORK CITED REFERENCES

Acharya, Sripathi, H. "Cost analysis and problems involved in processing and distribution of milk—A case study." Diss. U. University of Agricultural Sciences, Bangalore U, 1984.

Anand, S.K. "The ratio analysis aspects of a State Co-operative Consumers' Federation." *Indian Co-operative Review* 18(3) (1981): 263-267.

Ananth, G.S. "Performance Appraisal of the Bangalore Grape Growers' Marketing and Processing Co-operative Society Ltd"., Diss. Bangalore U, 1984.

Anonymous. "Company affairs-Eid-Parry (India) Ltd., *Indian Sugar,* 24(2) (1979): 115.

Ananymous. "Co-operative sugar factories make great stride", *Agriculture Situation in India,* 39(3)(1984): 167.

Athare, S.V. "Economic Evaluation of Maharashtra co-operative sugar industry with reference to its crushing capacity and use of by-product". Diss. Mahatma Phule Agricultural University, Rahuri U, 1982.

Babureddy, D.R and Chengappa, P.G. "Operational growth of Primary Co-operative Agriculture and Rural Development Banks in Karnataka-An Analysis". *Indian Co-operative Review* 339(2) (2004): 129-138.

Bajpai, S.K, ed., Prospect of reduction in cost of production of sugar in India, New Delhi: Seminar Report - 42nd *Annual Convention of Sugar Technologist Association of India,* 42 (3) (1978): 89-91.

Baviskar, B.S. Dairy Development in a tribal area of Gujarat, *Working Paper Series,* Institute of Social Studies, Netherlands, (1980) No. 8:47.

Bhalerao, M.M. and Vishwanath. "Co-operative Marketing of Fruits and Vegetables". *Khadhigramodyoga,* 18(6) (1972): 365-369.

Bhalerao, M.M., Singh, R.K. and Singh, T.N. "Growth of Areca nut Marketing societies in India". *Indian Co-operative Review,* 19(1) (1981); 19-32.

Chougule, J.D. "High sugar recovery performance of sugar factories in Andhra Pradesh", *Bhartiya Sugar,* 11(9) (1986) : 45.

Dhankar, B.L. "A Co-operative Primary Bank- Promise and Performance". *Indian Co-operative Review,* 9(1) (1971): 61-80.

Government of India. Tariff commission. Report on cost structure of and the fair price payable to the sugar Industry. New Delhi: GOI, 1973.

Government of India. Report of the Committee constituted to examine the matters relating to economic viability of new sugar factories. New Delhi: GOI, 1974.

Government of India. Sugar Enquiry Commission Report, New Delhi: GOI, 1974.

Gajja, B.L. "Cost analysis of milk processing in aried zone of Rajasthan". *Indian Co-operative Review,* 23(1) (1985): 82-93.

Gracy, C.P. "An analysis of performance of Primary Agricultural Societies in different states of India".Diss.U, University of Agricultural Science, Bangalore U, 1983.

Guptha,V.K. and George, P.S. eds., Modernization of rice processing industry in Punjab. Ahmedabad: *Seminar Report-Indian Institute f Management,* 1974.

Guptha, V.K., Gopalaswamy, T.P. and Mathur, D.P. eds., Setting up of modern rice mills. Ahemadabad: *Seminar Report-Indian Institute of Management,* 1971

Guptha Yogesh. "Performance of Agricultural Credit Co-operative Societies in Shimla District of Himachal Pradesh". *Indian Co-operative Review,* 339(1) (2003): 241- 243.

Hegde, Dayananda, S. "Performance of Karnataka Oil Seeds Growers Federation Ltd., An Economic Analysis." Diss. U. University of Agricultural Sciences, Bangalore U, 1993.

Hereker, P.M. "Correlation analysis of financial and operational actors of Sugar Industry". *Indian Co-operative Review,* 33(1) (1995): 33-48.

Hind, A.M. "Co-operatives under performers nature of an exploratory analysis of co-operative and non co-operative companies in the agribusiness sector". *Journal of Agricultural Economics,* 45(2) (1994): 213-219

Hirunandez, J., Denabax, F. and Corina, M. "Study on economics of scale in raw sugar factories". *The International Sugar Journal,* (1971) 211.

Jain, S.L. "Sugar Development Fund", *Indian Sugar,* 37(1) (1987): 2.

Jawanram. "Organization and working of dairy co-operatives in Rajasthan: A case Study". *Indian Co-operative Review,* 25(3) (1993): 273-282.

Kamat,, G.S. "Management of Co-operative Sugar Factories in Maharashtra". *The Maharashtra Co-operative Quarterly,* LIX (3) (1975): 175.

Karche, L.M. "A Co-operative sugar industry in Marathwada, lead industry for development - Reference to sick factories from Marathwada and Vidharbha". *The Maharashtra Co-operative Quarterly,* 2(5) (1977): 15-20.

Karache, L.M. "The cane development in the sugar factory area with special reference to sick factories from Marathwada and Vidharbha". *The Maharashtra Co-operative Quarterly,* 3 (5) (1978):17-26.

Kasar, D.V., and Tilekar, S.N. "Impact of sugar industry on employment and income of seasonal migratory from house holds in Maharashtra". *Indian Journal of Agricultural Economics,* 44 (3) (1989): 329.

Kharkar, N.C. eds., Technological evaluation of sugar factories in Maharashtra. Pune: Seminar Report-*Deccan Sugar Technologists Association's,* 1975, pp.15-18.

Krishnaswami, O.R. and Narayanaswamy. N. "Determinants of capital structure". *The Tamil Nadu Journal of Co-operation,* 82 (5) (1990): 267-275.

Kumar, B. "Accounting for profitability". The cooperator, 20 (3) (1983): 621-622.

Mahalingam, S. "Sugar mills and cane development activities in Dharmaphuri District of Tamil Nadu". *Indian Co-operative Review,* 17(2) (1980): 107-113.

Mane, P.M. "Technological Socio-economic and political impact of sugarcane co-operative of rural Maharashtra". *Maharashtra Sugar,* 6(4) (1981): 47-48.

Mariyanthan, A.A. "primary Agricultural Credit Societies- their progress, performance and problems". *Indian Co-operative Review,* 9 (1973): 465-471.

Maruthi Pillai, B. "Crisis in South Sugar Mills". *Indian Sugar,* 20(1) (1971): 713-717.

Mattigatti, R.M. "Performance of Milk Producers' co-operative societies and their impact on dairy farming in Dharwad District, Karnataka." Diss. U. University of Agricultural Sciences, Dharwad U, 2000.

Greener, Michael. Between the lines of balance sheet. Oxford :*Pergamon International Library,* 1980 pp. 15-82.

Mittal, B.L. "Rehabilization and modernization of sugar factories Sugar Technology". *Indian Sugar,* 26 (5) (1976): 281.

Mohammad, H. and Raghuram, P. "Cashew processing and marketing". *Agricultural Marketing,* 30 (3) (1990): 187-189.

Mohandas, M. "A study on the effect of loyalty of primary societies on the viability of the Kerala State Co-operative Rubber Marketing federation Limited". *Indian Co-operative Review,* 33 (1) (1991): 126-134.

Mohite, V.J. Productivity of sugarcane changing patterns, sugarcane industry of Maharashtra-A blue print for progress. Bombay: Government of Maharashtra, 1974.

Moni, S. "Prospects and problems of the Tamil Nadu Sugar Industry". *SISSTA Journal,* 1 (2) (1975): 17-20.

Muralidharan, C.R. "Competitive study of processing of sugarcane into sugar, gur and Khandasari- A Case study in Mandya taluk, Mandya district." Diss. U. University of Agricultural Sciences, Bangalore U, 1981

Naidu Mohan, K. and Shrinivasan, T.R. "Sugarcane breeding", *Institute History Achievements,* 1 (1987): 1.

Narasaiah, A. "Cash management in a co-operative sugar factory- A case study". *Indian Co-operative Review,* 33(1) (1996): 268-214.

Narayanaswamy, N. and Ramachandran, N.S.R. "Factors affecting the growth and development of co-operative sugar factories—A case study". *Indian Co-operative Review,* 25 (1988): 243-245.

Nikam, G.A. "Financial strength of sugar co-operative- A radio analysis approach". *Indian Co-operative Review,* 23 (3) (1986): 244-254.

Padmini, E.V.K., and Lekha, P.M. Financial performance of industrial co-operatives- A case study of Sree Narayana Power Loom Industrial Co-Operatives Society. *Co- operative Perspective,* 26 (4) (1992): 93-105.

Pai, V S., Vadivel, V. and Kamale, K.H. "Diversified companies and financial performance- A study". *Finance India,* 9 (4) (1995): 977-988.

Pandey, I.M., and Bhat, R. Financial ratio patterns in Indian manufacturing Companies: A multivariate analysis. Ahemadabad. Indian Institute of Management, 1988.

Parthasarathy, P.B. "Regional variations in resources productivity of Sugarcane farms in Andhra Pradesh". *Indian Journal of Agricultural Economics,* 29 (3) (1974): 188-189.

Pawar, J.R. "Co-operative sugar units in Maharashtra economics of scale".The Economic *Times,* 16 (330) (1977): 5-8.

Phadnis, D. "Sugar Co-operative with the message". *The Maharashtra Co-operative Quarterly,* 60 (2) (1976): 95.

Muralidharan, C.R. "Competitive study of processing of sugarcane into sugar, gur and Khandasari—A Case study in Mandya taluk, Mandya district." Diss. U. University of Agricultural Sciences, Bangalore U, 1981

Pradeep Kumar, G.S. "An evaluation of business performance of the Horticultural Producers Co-operative Marketing and Processing Society Limited." Diss. U. University of Agricultural Sciences, Bangalore U. 2004.

Ramesh, N.A. "Adjudication of the performance of sugar factories". *Maharashtra Sugar,* 6 (4) (1980): 47-48.

Rao Desinga, N. "Co-operative credit movement in Andhra Pradesh after 1956". *Indian Co-operative Review,* 14 (3) (1977): 239-240.

Rayudu, C.S. "Radio analysis and financial performance". *Indian Co-operative Review,* 23 (1) (1985): 54-70.

Shah, H.P. and Shah, N.K. Measures for reduction in cost of production of sugar, *Gujarat Association for Agricultural Science,* 1980.

Shankaramurthy, H.G. "Performance of the Karnataka State Co-operative Marketing Federation Ltd. and its impact on farm market—An economic Analysis". Diss. U. Andhra Pradesh Agricultural University, Hyderabad U, 1986.

Shindhe Jayajirao. "Editorial sick sugar factories". *Maharashtra Sugar,* 6 (7) (1981): 5.

Srivatsava, U.K. "Agro processing industries- potential, Constraints and task Ahead". *Indian Journal of Agricultural Economics,* 44 (3) (1995): 12-15.

Srvatsava, U.K., and Mukunda Das, V. The economics of sunflower cultivation and processing with special reference to Tamil Nadu. Ahmedabad. Indian Institute of Management Publication, 1975

Subbarao, A.R. "An analysis of business performance of the Central Areca nut Marketing and Processing Co-operative Ltd., Mangalore, Karnataka". Diss. U. University of Agricultural Sciences, Bangalore U. 2000

Tewari, S.C. and George, M.V. "Marketing of agricultural products through co-operative marketing societies-*An opinion survey". Indian Co-operative Review,* 8 (1970): 206-214.

Thanulingam, N. and Gurumoorthy, TR. "Financial performance of Handloom co-operative societies—A case study". *Indian Co-operative Review,* 25 (2) (1987): 132-142.

Thaper, B.M. Way in sugar industry sick, Financial Express: 1976.

GOVERNMENT REPORT *Government of India. Tariff commission. Report on cost structure of and the fair price payable to the sugar Industry. New Delhi: GOI, 1973.

Tuteja. Sugar Commission. Revitalization of sick sugar Industry in India. New Delhi: Government of India. 2004

Venkatesan, K. "Modernization and improvement in productivity". *SISSTA Sugar Journal* 13 (4) (1987): 7.

Verma, N.K. "Growth of dairy Co-operatives in India". Indian Co-operative Review, 18 (1) (1980): 45-49.

APPENDICES

APPENDIX - 1: SECTOR WISE SUGAR PRODUCTION IN KARNATAKA FROM 1990-91 TO 2004-05

In Metric Tonnes

YEAR	*Sugar Factories Under*							
	Co-operative Sector		*Public Sector*		*Private Sector*		*Over All*	
	Quantity	*Percentage*	*Quantity*	*Percentage*	*Quantity*	*Percentage*	*Quantity*	*Percentage*
1990 - 91	451822.00	48.12	131863.00	14.04	355298.00	37.84	938983.00	100.00
1991 - 92	497413.00	48.00	141906.00	13.69	397028.00	38.31	1036347.00	100.00
1992 - 93	402909.00	47.48	127977.00	15.08	317645.00	37.43	848531.00	100.00
1993 - 94	375500.00	45.16	109010.00	13.11	347000.00	41.73	831510.00	100.00
1994 - 95	569000.00	46.37	159900.00	13.03	498300.00	40.60	1227200.00	100.00
1995 - 96	614500.00	48.33	161660.00	12.71	495397.00	38.96	1271557.00	100.00
1996 - 97	421800.00	47.99	122800.00	13.97	334300.00	38.04	878900.00	100.00
1997 - 98	466600.00	48.04	86700.00	8.93	418000.00	43.04	971300.00	100.00
1998 - 99	712500.00	52.27	120800.00	8.86	529900.00	38.87	1363200.00	100.00
1999 - 00	756600.00	47.56	147500.00	9.27	686700.00	43.17	1590800.00	100.00
2000 - 01	673400.00	41.57	142900.00	8.82	803700.00	49.61	1620000.00	100.00
2001 - 02	549814.00	35.32	125169.00	8.04	881678.00	56.64	1556661.00	100.00
2002 - 03	658355.00	35.01	115909.00	6.16	1106132.00	58.82	1880396.00	100.00
2003 - 04	328282.00	29.20	61741.00	5.49	734166.00	65.31	1124189.00	100.00
2004 - 05	525343.00	33.97	19270.00	1.25	1001840.00	64.78	1546453.00	100.00
AVG		43.63		10.16		46.21		100.00
CGR	1.581237		-4.06263		8.494949		4.340655	

Source: *Field Survey and Analyses of Sample Sugar Factories (Refer Nature and Source of Data Collection).*

APPENDIX - 2: QUANTITY OF SUGARCANE CRUSHED IN KARNATAKA-SECTOR WISE FROM 1990-91 TO 2004-05

YEAR	*Sugar Factories Under*						*Over All Industry*	
	Co-operative Sector		*Public Sector*		*Private Sector*			
	Quantity	*Percentage*	*Quantity*	*Percentage*	*Quantity*	*Percentage*	*Quantity*	*Percentage*
1990 - 91	4409378.00	48.11	1346555.00	14.69	3409956.00	37.20	9165889.00	100.00
1991 - 92	4703133.00	47.33	1487223.00	14.97	3746024.00	37.70	9936380.00	100.00
1992 - 93	3772365.00	47.13	1344827.00	16.80	2887259.00	36.07	8004451.00	100.00
1993 - 94	3530213.00	44.50	1130067.00	14.25	3272777.00	41.25	7933057.00	100.00
1994 - 95	5559850.00	46.82	1663708.00	14.01	4651660.00	39.17	11875218.00	100.00
1995 - 96	6143923.00	47.64	1783680.00	13.83	4968666.00	38.53	12896269.00	100.00
1996 - 97	3872336.00	46.84	1301889.00	15.75	3092418.00	37.41	8266643.00	100.00
1997 - 98	4324480.00	47.35	943628.00	10.33	3864622.00	42.32	9132730.00	100.00
1998 - 99	6647690.00	51.09	1389201.00	10.68	4975552.00	38.24	13012443.00	100.00
1999 - 00	6977700.00	46.90	1516069.00	10.19	6383340.00	42.91	14877109.00	100.00
2000 - 01	6154270.00	41.19	1449668.00	9.70	7337194.00	49.11	14941132.00	100.00
2001 - 02	5019997.00	34.68	1343524.00	9.28	8112111.00	56.04	14475632.00	100.00
2002 - 03	5993756.00	34.60	1204103.00	6.95	10125405.00	58.45	17323264.00	100.00
2003 - 04	3079463.00	29.01	667736.00	6.29	6867616.00	64.70	10614815.00	100.00
2004 - 05	58403340.00	83.63	1193440.00	1.71	10234560.00	14.66	69831340.00	100.00
AVG		46.46		11.30		42.25		100.00
CGR	16.31024		-1.8779		8.605403		11.97871	

Source: *Field Survey and Analyses of Sample Sugar Factories (Refer Nature and Source of Data Collection)*

APPENDIX - 3: DURATION OF SUGARCANE CRUSHED IN KARNATAKA – SECTOR WISE FROM 1990-91 TO 2004-05

YEAR	SUGAR FACTORIES UNDER CO-OPERATIVE SECTOR	PUBLIC SECTOR	PRIVATE SECTOR	OVER ALL INDUSTRY
1990 - 91	157.00	132.00	147.00	145.33
1991 - 92	165.00	147.00	155.00	155.67
1992 - 93	150.00	152.00	138.00	146.67
1993 - 94	112.00	134.00	135.00	127.00
1994 - 95	169.00	183.00	192.00	181.33
1995 - 96	213.00	182.00	201.00	198.67
1996 - 97	115.00	134.00	141.00	130.00
1997 - 98	110.00	133.00	135.00	126.00
1998 - 99	159.00	178.00	175.00	170.67
1999 - 00	165.00	182.00	176.00	174.33
2000 - 01	168.00	150.00	160.00	159.33
2001 - 02	156.00	133.00	147.00	145.33
2002 - 03	164.00	152.00	161.00	159.00
2003 - 04	155.00	134.00	145.00	144.67
2004 - 05	165.00	153.00	161.00	159.67
AVG	154.87	151.93	157.93	154.91

***Source:** Field Survey and Analyses of Sample Sugar Factories (Refer Nature and Source of Data Collection)*

APPENDIX– 4: RECOVERY PERCENTAGES OF SUGAR FROM SUGARCANE IN KARNATAKA– SECTOR WISE FROM 1990-91 TO 2004-05

YEAR	*SUGAR FACTORIES UNDER*			*OVER ALL*
	CO-OPERATIVE SECTOR	*PUBLIC SECTOR*	*PRIVATE SECTOR*	*INDUSTRY*
1990 - 91	10.12	10.11	9.78	10.00
1991 - 92	10.47	9.81	9.98	10.11
1992 - 93	10.24	9.76	10.37	10.31
1993 - 94	10.62	10.11	10.73	10.40
1994 - 95	10.58	10.01	10.32	10.38
1995 - 96	10.94	10.36	10.06	10.52
1996 - 97	10.63	10.21	10.89	10.56
1997 - 98	10.59	10.18	10.88	10.55
1998 - 99	10.58	10.19	10.89	10.60
1999 - 00	10.83	10.24	10.85	10.54
2000 - 01	10.93	10.35	10.01	10.58
2001 - 02	10.99	10.33	10.89	10.79
2002 - 03	10.81	10.74	10.97	10.58
2003 - 04	10.56	10.05	10.33	10.45
2004 - 05	10.60	10.24	10.89	10.51
AVG	10.63	10.18	10.52	10.46

Source: *Field Survey and Analyses of Sample Sugar Factories (Refer Nature and Source of Data Collection)*

APPENDIX - 5: INSTALLED CAPACITY OF SELECTED SUGAR FACTORIES FROM 1990-91 TO 2004-05

YEARS	CO-OPERATIVE SECTOR								
	ABOVE AVERAGE				BELOW AVERAGE				AVG
	HIRANYA-KESHI	MALA-PRABHA	DOOD-GHANGA	AVG	PANDAVA-PURA	GHATA-PRABHA	HEMA-VATHI	AVG	
1990 - 91	5000.00	3500.00	2000.00	3500.00	1500.00	1250.00	1250.00	1333.33	2416.67
1991 - 92	5000.00	3500.00	2000.00	3500.00	1500.00	1250.00	1250.00	1333.33	2416.67
1992 - 93	5000.00	3500.00	2000.00	3500.00	1500.00	1250.00	1250.00	1333.33	2416.67
1993 - 94	5000.00	3500.00	2000.00	3500.00	1500.00	1250.00	1250.00	1333.33	2416.67
1994 - 95	5000.00	3500.00	2000.00	3500.00	1500.00	1250.00	1250.00	1333.33	2416.67
1995 - 96	5000.00	3500.00	3500.00	4000.00	1500.00	1250.00	1250.00	1333.33	2666.67
1996 - 97	5000.00	3500.00	3500.00	4000.00	1500.00	1250.00	1250.00	1333.33	2666.67
1997 - 98	5000.00	3500.00	3500.00	4000.00	1500.00	1250.00	1250.00	1333.33	2666.67
1998 - 99	5000.00	3500.00	3500.00	4000.00	3500.00	1250.00	1250.00	2000.00	3000.00
1999 - 00	5000.00	3500.00	3500.00	4000.00	3500.00	1250.00	1250.00	2000.00	3000.00
2000 - 01	5000.00	3500.00	3500.00	4000.00	3500.00	1250.00	1250.00	2000.00	3000.00
2001 - 02	5000.00	3500.00	3500.00	4000.00	3500.00	1250.00	1250.00	2000.00	3000.00
2002 - 03	5000.00	3500.00	3500.00	4000.00	3500.00	1250.00	1250.00	2000.00	3000.00
2003 - 04	5000.00	3500.00	3500.00	4000.00	3500.00	2200.00	1250.00	2316.67	3158.33
2004 - 05	5000.00	3500.00	3500.00	4000.00	3500.00	2200.00	1250.00	2316.67	3158.33
CGR	0.00	0.00	13.55	1.16	20.50	3.20	0.00	4.82	2.28

Source: *Field Survey and Analyses of Sample Sugar Factories (Refer Nature and Source of Data Collection)*

APPENDIX - 5a: INSTALLED CAPACITY OF SELECTED SUGAR FACTORIES FROM 1990-91 TO 2004-05

YEARS	*PUBLIC SECTOR*			*PRIVATE SECTOR*			*OVERALL AVG*
	ABOVE AVG	*BELOW AVG*	*AVG*	*ABOVE AVG*	*BELOW AVG*	*AVG*	*(Co-OPERATIVE,*
	MY SUGAR	*MYSORE*		*UGAR*	*DHAVANAGERE*		*PUBLIC AND PRIVATE)*
1990 - 91	5000.00	2500.00	3750.00	5000.00	1250.00	3125.00	3097.22
1991 - 92	5000.00	2500.00	3750.00	5000.00	1250.00	3125.00	3097.22
1992 - 93	5000.00	2500.00	3750.00	5000.00	1250.00	3125.00	3097.22
1993 - 94	5000.00	2500.00	3750.00	6500.00	1250.00	3875.00	3347.22
1994 - 95	5000.00	2500.00	3750.00	6500.00	1250.00	3875.00	3347.22
1995 - 96	5000.00	2500.00	3750.00	6500.00	1250.00	3875.00	3430.56
1996 - 97	5000.00	2500.00	3750.00	6500.00	1250.00	3875.00	3430.56
1997 - 98	5000.00	2500.00	3750.00	7500.00	1250.00	4375.00	3597.22
1998 - 99	5000.00	2500.00	3750.00	7500.00	1250.00	4375.00	3708.33
1999 - 00	5000.00	2500.00	3750.00	7500.00	1250.00	4375.00	3708.33
2000 - 01	5000.00	2500.00	3750.00	7500.00	1250.00	4375.00	3708.33
2001 - 02	5000.00	2500.00	3750.00	10000.00	1250.00	5625.00	4125.00
2002 - 03	5000.00	2500.00	3750.00	10000.00	1250.00	5625.00	4125.00
2003 - 04	5000.00	2500.00	3750.00	10000.00	1250.00	5625.00	4177.78
2004 - 05	5000.00	2500.00	3750.00	10000.00	1250.00	5625.00	4177.78
CGR	0.00	0.00	0.00	5.31	0.00	4.54	2.39

Source: *Field Survey and Analyses of Sample Sugar Factories (Refer Nature and Source of Data Collection)*

APPENDIX - 6: QUANTITY OF SUGARCANE CRUSHED OF SELECTED SUGAR FACTORIES FROM 1990-91 TO 2004-05

YEARS	*CO-OPERATIVE SECTOR*								
	ABOVE AVERAGE				*ABOVE AVERAGE*				*AVG*
	HIRANYA-KESHI	*MALA-PRABHA*	*DOOD-GHANGA*	*AVG*	*PANDAVA-PURA*	*GHATA-PRABHA*	*HEMA-VATHI*	*AVG*	
1990 - 91	812158.00	605248.00	436972.60	618126.20	283297.00	281040.59	148696.90	237678.16	427902.18
1991 - 92	748594.00	715502.00	448188.44	637428.15	246474.00	288825.00	294880.00	276726.33	457077.24
1992 - 93	753350.00	594363.00	440606.18	596106.39	223459.00	224181.00	234652.50	227430.83	411768.61
1993 - 94	643147.00	564959.00	333738.96	513948.32	169302.00	188898.00	154583.60	170927.87	342438.09
1994 - 95	946037.00	751223.00	344220.00	680493.33	274896.00	362018.00	340346.92	325753.64	503123.49
1995 - 96	1020668.00	821153.00	523473.00	788431.33	332323.00	393978.00	319250.92	348517.31	568474.32
1996 - 97	696997.00	497147.00	496802.60	563648.87	116769.00	223034.00	142531.40	160778.13	362213.50
1997 - 98	949822.00	425829.00	533030.00	636227.00	0.00	235895.00	82671.95	106188.98	371207.99
1998 - 99	1166826.00	782248.00	554361.00	834478.33	296737.00	345055.00	249690.80	297160.93	565819.63
1999 - 00	1054974.00	590012.00	764980.00	803322.00	331336.00	368242.00	25794.77	241790.92	522556.46
2000 - 01	1011152.00	621474.00	781986.51	804870.84	339850.00	405664.00	28339.68	257951.23	531411.03
2001 - 02	918296.00	414335.00	612415.48	648348.83	155721.00	377782.00	191007.00	241503.33	444926.08
2002 - 03	914223.00	501367.00	660312.89	691967.63	148012.00	448933.00	252273.50	283072.83	487520.23
2003 - 04	402792.00	244687.00	395310.69	347596.56	0.00	294150.00	167476.70	153875.57	250736.07
2004 - 05	509158.00	483436.00	494488.85	495694.28	0.00	456371.00	78325.00	178232.00	336963.14
CGR	-0.66	-3.30	2.84	-0.50	-6.46	3.66	-4.82	-1.33	-0.72
t Value	-2.074	1.823	-3.957	-1.394	-3.224	-2.343	-0.397	-2.783	-2.018

Source: *Field Survey and Analyses of Sample Sugar Factories (Refer Nature and Source of Data Collection)*

APPENDIX - 6a: QUANTITY OF SUGARCANE CRUSHED OF SELECTED SUGAR FACTORIES FROM 1990-91 TO 2004-05

YEARS	*PUBLIC SECTOR*			*PRIVATE SECTOR*			*OVERALL AVG (CO-OPERATIVE, PUBLIC AND PRIVATE)*
	ABOVE AVG	*BELOW AVG*	*AVG*	*ABOVE AVG*	*BELOW AVG*	*AVG*	
	MY SUGAR	*MYSORE*		*UGAR*	*DHAVANAGERE*		
1990 - 91	624654.00	371585.00	498119.50	1038576.00	183751.00	611163.50	512395.06
1991 - 92	664329.00	471110.00	567719.50	1015853.00	277080.00	646466.50	557087.75
1992 - 93	591229.00	374863.00	483046.00	718902.00	211790.00	465346.00	453386.87
1993 - 94	472561.00	251121.00	361841.00	680272.00	158859.00	419565.50	374614.86
1994 - 95	879147.00	419678.00	649412.50	1138830.00	276724.00	707777.00	620104.33
1995 - 96	779044.00	459611.00	619327.50	1393789.00	263534.00	828661.50	672154.44
1996 - 97	643151.00	671952.00	657551.50	947339.00	237366.00	592352.50	537372.50
1997 - 98	356818.00	377962.00	367390.00	1130355.00	216951.00	673653.00	470750.33
1998 - 99	840639.00	460356.00	650497.50	1399602.00	242979.00	821290.50	679202.54
1999 - 00	958135.00	503705.00	730920.00	1425023.00	320009.00	872516.00	708664.15
2000 - 01	889750.00	423782.00	656766.00	1189443.00	397050.00	793246.50	660474.51
2001 - 02	849919.00	396706.00	623312.50	1222415.00	421513.00	821964.00	630067.53
2002 - 03	899782.00	526676.00	713229.00	1467797.00	549512.00	1008654.50	736467.91
2003 - 04	294588.00	426341.00	360464.50	820996.00	344865.00	582930.50	398043.69
2004 - 05	706920.00	503734.00	605327.00	1635573.00	157056.00	896314.50	612868.21
CGR	1.02	1.52	1.14	2.98	4.33	3.25	1.57
t Value	-4.979	-3.484	-1.65	-4.36	-4.937	-6.832	-5.506

Source: *Field Survey and Analyses of Sample Sugar Factories (Refer Nature and Source of Data Collection)*

APPENDIX - 7: RECOVERY PERCENTAGE OF SUGARCANE OF SELECTED SUGAR FACTORIES FROM 1990-91 TO 2004-05

YEARS	*CO-OPERATIVE SECTOR*								
	ABOVE AVERAGE				*ABOVE AVERAGE*				*AVG*
	HIRANYA-KESHI	*MALA-PRABHA*	*DOOD-GHANGA*	*AVG*	*PANDAVA-PURA*	*GHATA-PRABHA*	*HEMA-VATHI*	*AVG*	
1990 - 91	10.00	10.91	10.56	10.49	9.70	10.69	9.91	10.10	10.30
1991 - 92	11.47	11.26	11.56	11.43	9.12	11.22	9.77	10.04	10.73
1992 - 93	10.98	11.55	11.66	11.40	9.13	11.42	9.56	10.04	10.72
1993 - 94	10.93	11.41	12.49	11.61	9.45	11.23	8.72	9.80	10.71
1994 - 95	10.42	10.96	12.55	11.31	9.10	11.17	9.50	9.92	10.62
1995 - 96	9.83	10.35	11.41	10.53	9.13	10.67	8.80	9.53	10.03
1996 - 97	11.86	11.05	10.86	11.26	9.00	10.83	8.54	9.46	10.36
1997 - 98	11.42	11.07	11.59	11.36	0.00	10.77	8.23	6.33	8.85
1998 - 99	11.58	10.61	11.69	11.29	6.69	11.28	9.50	9.16	10.23
1999 - 00	11.72	10.64	11.75	11.37	8.55	11.61	9.30	9.82	10.60
2000 - 01	11.79	10.54	11.56	11.30	8.90	11.59	9.30	9.93	10.61
2001 - 02	11.71	10.37	11.79	11.29	8.80	11.10	9.00	9.63	10.46
2002 - 03	11.32	10.80	11.90	11.34	8.92	11.20	8.30	9.47	10.41
2003 - 04	11.23	9.84	10.20	10.42	0.00	10.33	10.12	6.82	8.62
2004 - 05	10.16	10.08	10.81	10.35	0.00	10.54	7.90	6.15	8.25
AVG	11.09	10.76	11.49	11.12	7.10	11.04	9.10	9.08	10.10

Source: *Field Survey and Analyses of Sample Sugar Factories (Refer Nature and Source of Data Collection)*

APPENDIX - 7a: RECOVERY PERCENTAGE OF SUGARCANE OF SELECTED SUGAR FACTORIES FROM 1990-91 TO 2004-05

YEARS	PUBLIC SECTOR			PRIVATE SECTOR			OVERALL AVG (CO-OPERATIVE, PUBLIC AND PRIVATE)
	ABOVE AVG	BELOW AVG	AVG	ABOVE AVG	BELOW AVG	AVG	
	MY SUGAR	MYSORE		UGAR	DHAVANAGERE		
1990 - 91	9.65	10.20	9.93	11.32	10.11	10.72	10.31
1991 - 92	9.21	10.31	9.76	12.18	9.59	10.89	10.46
1992 - 93	8.84	9.95	9.40	12.74	9.55	11.15	10.42
1993 - 94	9.00	10.11	9.56	12.58	8.94	10.76	10.34
1994 - 95	9.40	10.33	9.87	12.27	9.77	11.02	10.50
1995 - 96	9.30	10.18	9.74	10.90	9.44	10.17	9.98
1996 - 97	7.83	9.28	8.56	11.51	10.29	10.90	9.94
1997 - 98	8.25	9.54	8.90	11.45	10.36	10.91	9.55
1998 - 99	8.84	10.03	9.44	10.92	10.38	10.65	10.10
1999 - 00	9.28	10.15	9.72	11.84	10.38	11.11	10.47
2000 - 01	9.31	10.48	9.90	12.10	10.44	11.27	10.59
2001 - 02	9.54	10.37	9.96	11.59	10.18	10.89	10.43
2002 - 03	9.10	10.31	9.71	11.57	10.22	10.90	10.34
2003 - 04	8.75	9.45	9.10	10.13	8.73	9.43	9.05
2004 - 05	5.36	10.34	7.85	11.53	8.64	10.09	8.73
AVG	8.78	10.07	9.42	11.64	9.80	10.72	10.08

Source: *Field Survey and Analyses of Sample Sugar Factories (Refer Nature and Source of Data Collection)*

APPENDIX - 8: DURATION OF CRUSHING SEASON OF SELECTED SUGAR FACTORIES FROM 1990-91 TO 2004-05

YEARS	CO-OPERATIVE SECTOR								
	ABOVE AVERAGE				ABOVE AVERAGE				AVG
	HIRANYA-KESHI	MALA-PRABHA	DOOD-GHANGA	AVG	PANDAVA-PURA	GHATA-PRABHA	HEMA-VATHI	AVG	
1990 - 91	163.00	172.00	218.00	184.33	268.00	217.00	131.00	205.33	194.83
1991 - 92	162.00	212.00	241.00	205.00	241.00	222.00	248.00	237.00	221.00
1992 - 93	156.00	186.00	210.00	184.00	218.00	159.00	207.00	194.67	189.33
1993 - 94	132.00	166.00	158.00	152.00	168.00	124.00	179.00	157.00	154.50
1994 - 95	199.00	207.00	156.00	187.33	252.00	229.00	301.00	260.67	224.00
1995 - 96	223.00	255.00	218.00	232.00	335.00	245.00	319.00	299.67	265.83
1996 - 97	154.00	164.00	226.00	181.33	124.00	162.00	155.00	147.00	164.17
1997 - 98	192.00	124.00	166.00	160.67	0.00	149.00	98.00	82.33	121.50
1998 - 99	232.00	217.00	168.00	205.67	296.00	215.00	231.00	247.33	226.50
1999 - 00	220.00	168.00	214.00	200.67	214.00	220.00	234.00	222.67	211.67
2000 - 01	215.00	171.00	212.00	199.33	214.00	225.00	248.00	229.00	214.17
2001 - 02	184.00	122.00	151.00	152.33	142.00	203.00	233.00	192.67	172.50
2002 - 03	179.00	135.00	161.00	158.33	122.00	208.00	236.00	188.67	173.50
2003 - 04	91.00	71.00	109.00	90.33	0.00	138.00	136.73	91.58	90.96
2004 - 05	86.00	168.00	128.00	127.33	0.00	201.00	100.00	100.33	113.83
AVG	172.53	169.20	182.40	174.71	172.93	194.47	203.78	190.39	182.55

Source: *Field Survey and Analyses of Sample Sugar Factories (Refer Nature and Source of Data Collection)*

APPENDIX - 8a: DURATION OF CRUSHING SEASON OF SELECTED SUGAR FACTORIES FROM 1990-91 TO 2004-05

YEARS	*PUBLIC SECTOR*			*PRIVATE SECTOR*			*OVERALL AVG (CO-OPERATIVE, PUBLIC AND PRIVATE)*
	ABOVE AVG	*BELOW AVG*	*AVG*	*ABOVE AVG*	*BELOW AVG*	*AVG*	
	MY SUGAR	*MYSORE*		*UGAR*	*DHAVANAGERE*		
1990 - 91	281.00	159.00	220.00	254.00	144.00	199.00	204.61
1991 - 92	269.00	198.00	233.50	247.00	200.00	223.50	226.00
1992 - 93	238.00	208.00	223.00	158.00	174.00	166.00	192.78
1993 - 94	223.00	124.00	173.50	164.00	150.00	157.00	161.67
1994 - 95	287.00	190.00	238.50	248.00	231.00	239.50	234.00
1995 - 96	266.00	214.00	240.00	310.00	242.00	276.00	260.61
1996 - 97	336.00	229.00	282.50	192.00	162.00	177.00	207.89
1997 - 98	174.00	167.00	170.50	202.00	189.00	195.50	162.50
1998 - 99	279.00	209.00	244.00	267.00	186.00	226.50	232.33
1999 - 00	258.00	221.00	239.50	258.00	290.00	274.00	241.72
2000 - 01	256.00	189.00	222.50	203.00	188.00	195.50	210.72
2001 - 02	229.00	157.00	193.00	176.00	207.00	191.50	185.67
2002 - 03	252.00	218.00	235.00	195.00	230.00	212.50	207.00
2003 - 04	114.00	183.00	148.50	116.00	134.00	125.00	121.49
2004 - 05	244.00	213.00	228.50	205.00	94.00	149.50	163.94
AVG	247.07	191.93	219.50	213.00	188.07	200.53	200.86

Source: *Field Survey and Analyses of Sample Sugar Factories (Refer Nature and Source of Data Collection)*

APPENDIX - 9: SUGAR PRODUCTION OF SELECTED SUGAR FACTORIES FROM 1990-91 TO 2004-05

YEARS	CO-OPERATIVE SECTOR								
	ABOVE AVERAGE				ABOVE AVERAGE				AVG
	HIRANYA-KESHI	MALA-PRABHA	DOOD-GHANGA	AVG	PANDAVA-PURA	GHATA-PRABHA	HEMA-VATHI	AVG	
1990 - 91	793602.00	668690.00	483854.00	648715.33	274870.00	300030.00	141250.00	238716.67	443716.00
1991 - 92	868800.00	822665.00	497145.00	729536.67	225815.00	324430.00	288206.00	279483.67	504510.17
1992 - 93	834700.00	698460.00	919423.00	817527.67	202515.00	259470.00	223840.00	228608.33	523068.00
1993 - 94	725040.00	648170.00	429480.00	600896.67	160354.00	212940.00	134855.00	169383.00	385139.83
1994 - 95	994920.00	830440.00	429579.00	751646.33	251345.00	401550.00	322520.00	325138.33	538392.33
1995 - 96	1010200.00	859030.00	590522.00	819917.33	317105.00	425150.00	281560.00	341271.67	580594.50
1996 - 97	1031200.00	561380.00	556731.00	716437.00	106185.00	245570.00	121520.00	157758.33	437097.67
1997 - 98	1002650.00	486060.00	615951.00	701553.67	0.00	253730.00	71665.00	108465.00	405009.33
1998 - 99	1121430.00	834620.00	642283.00	866111.00	198340.00	389580.00	241890.00	276603.33	571357.17
1999 - 00	1300800.00	640360.00	897371.00	946177.00	286520.00	427670.00	238958.00	317716.00	631946.50
2000 - 01	1297790.00	661180.00	904275.00	954415.00	302690.00	470145.00	263360.00	345398.33	649906.67
2001 - 02	1317230.00	441200.00	749535.00	835988.33	134990.00	419440.00	170280.00	241570.00	538779.17
2002 - 03	987690.00	565015.00	800440.00	784381.67	132085.00	501620.00	222320.00	285341.67	534861.67
2003 - 04	553910.00	251875.00	416539.00	407441.33	0.00	302650.00	171940.00	158196.67	282819.00
2004 - 05	517450.00	226802.00	534334.00	426195.33	0.00	482740.00	62100.00	181613.33	303904.33
CGR	0.33	-5.06	1.52	-0.83	-7.33	3.50	-2.28	-0.62	-0.78
t Value	-5.617	-4.346	-5.866	-2.446	-0.899	-5.726	1.304	-2.152	-1.625

Source: *Field Survey and Analyses of Sample Sugar Factories (Refer Nature and Source of Data Collection)*

APPENDIX - 9a: SUGAR PRODUCTION OF SELECTED SUGAR FACTORIES FROM 1990-91 TO 2004-05

YEARS	*PUBLIC SECTOR*			*PRIVATE SECTOR*			*OVERALL AVG*
	ABOVE AVG	*BELOW AVG*	*AVG*	*ABOVE AVG*	*BELOW AVG*	*AVG*	*(CO-OPERATIVE, PUBLIC AND PRIVATE)*
	MY SUGAR	*MYSORE*		*UGAR*	*DHAVANAGERE*		
1990 - 91	602840.00	380405.00	491622.50	1176130.00	185807.00	680968.50	538769.00
1991 - 92	602890.00	486970.00	544930.00	1237118.00	265632.00	751375.00	600271.72
1992 - 93	52248.00	379945.00	216096.50	922037.00	202299.00	562168.00	433777.50
1993 - 94	42565.00	255980.00	149272.50	859300.00	141999.00	500649.50	345020.61
1994 - 95	82635.00	431530.00	257082.50	1385200.00	270402.00	827801.00	541091.94
1995 - 96	72422.00	469595.00	271008.50	1539680.00	252082.00	895881.00	582494.67
1996 - 97	50330.00	447480.00	248905.00	1089972.00	244198.00	667085.00	451029.22
1997 - 98	29403.00	365580.00	197491.50	1294251.00	224811.00	759531.00	454010.61
1998 - 99	74316.00	462540.00	268428.00	1552757.00	252188.00	902472.50	580752.56
1999 - 00	89417.00	514410.00	301913.50	1715195.00	332405.00	1023800.00	652553.33
2000 - 01	83774.00	463399.00	273586.50	1481160.00	416031.00	948595.50	624029.56
2001 - 02	81129.00	414890.00	248009.50	1433272.00	424335.00	928803.50	571864.06
2002 - 03	81846.00	558380.00	320113.00	1713725.00	561635.00	1137680.00	664218.22
2003 - 04	26754.00	404070.00	215412.00	871700.00	300923.00	586311.50	361514.17
2004 - 05	358810.00	408430.00	383620.00	1894534.00	135730.00	1015132.00	567552.11
CGR	-11.03	1.19	-2.06	2.71	4.32	2.99	0.91
t Value	2.519	-6.973	-3.297	-6.665	4.914	-4.173	-1.409

Source: *Field Survey and Analyses of Sample Sugar Factories (Refer Nature and Source of Data Collection)*

APPENDIX - 10: SHARE CAPITAL POSITION OF SELECTED SUGAR FACTORIES FROM 1990-91 TO 2004-05 (Rs in lakhs)

YEARS	CO-OPERATIVE SECTOR								
	ABOVE AVERAGE				*ABOVE AVERAGE*				*AVG*
	HIRANYA-KESHI	*MALA-PRABHA*	*DOOD-GHANGA*	*AVG*	*PANDAVA-PURA*	*GHATA-PRABHA*	*HEMA-VATHI*	*AVG*	
1990 - 91	144.05	192.76	425.03	253.95	412.00	402.94	300.00	371.65	312.80
1991 - 92	144.03	192.86	425.03	253.97	412.00	402.94	300.00	371.65	312.81
1992 - 93	145.67	193.08	1100.00	479.58	412.00	404.28	300.00	372.09	425.84
1993 - 94	145.94	193.08	1100.00	479.67	412.00	534.50	500.00	482.17	480.92
1994 - 95	146.27	240.12	1100.00	495.46	412.00	509.96	500.00	473.99	484.73
1995 - 96	146.76	313.07	1100.00	519.94	412.00	510.00	500.00	474.00	496.97
1996 - 97	212.26	313.18	1100.00	541.81	412.00	559.02	500.00	490.34	516.08
1997 - 98	241.92	313.16	1198.00	584.36	412.00	592.55	500.00	501.52	542.94
1998 - 99	242.07	365.88	1198.00	601.98	412.00	598.09	500.00	503.36	552.67
1999 - 00	242.32	365.87	1000.00	536.06	412.00	598.09	662.78	557.62	546.84
2000 - 01	242.56	367.22	1000.00	536.59	412.00	598.55	667.90	559.48	548.04
2001 - 02	274.13	750.26	1500.00	841.46	412.00	664.71	671.80	582.84	712.15
2002 - 03	279.17	800.25	1500.00	859.81	412.00	664.76	738.59	605.12	732.46
2003 - 04	789.28	800.66	1500.00	1029.98	412.00	664.80	797.62	624.81	827.39
2004 - 05	792.36	800.78	1111.19	901.44	0.00	664.84	819.06	494.63	698.04
CGR	12.92	11.84	4.73	7.66	-2.68	3.65	6.70	3.15	5.60
t Value	-9.233	-6.173	-4.964	-7.552	-1	-4.313	-5.571	-6.094	-7.453

Source: *Field Survey and Analyses of Sample Sugar Factories (Refer Nature and Source of Data Collection)*

APPENDIX - 10a: SHARE CAPITAL POSITION OF SELECTED SUGAR FACTORIES FROM 1990-91 TO 2004-05 (Rs in lakhs)

YEARS	*PUBLIC SECTOR*			*PRIVATE SECTOR*			*OVERALL AVG*
	ABOVE AVG	*BELOW AVG*	*AVG*	*ABOVE AVG*	*BELOW AVG*	*AVG*	*(CO-OPERATIVE, PUBLIC AND PRIVATE)*
	MY SUGAR	*MYSORE*		*UGAR*	*DHAVANAGERE*		
1990 - 91	300.48	1725.31	1012.89	196.00	201.99	198.99	508.23
1991 - 92	273.43	1613.25	943.34	196.00	217.93	206.96	487.70
1992 - 93	273.43	1613.25	943.34	200.00	235.69	217.84	529.01
1993 - 94	273.43	2463.25	1368.34	200.00	236.09	218.04	689.10
1994 - 95	273.43	2463.25	1368.34	363.30	238.44	300.87	717.98
1995 - 96	273.43	2463.25	1368.34	375.00	239.04	307.02	724.11
1996 - 97	273.43	2463.25	1368.34	375.00	239.13	307.06	730.49
1997 - 98	473.43	2463.25	1468.34	450.00	366.24	408.12	806.47
1998 - 99	673.43	2463.25	1568.34	561.56	366.25	463.91	861.64
1999 - 00	673.43	2936.54	1804.99	562.50	548.67	555.59	969.14
2000 - 01	673.43	2936.54	1804.99	562.50	708.01	635.26	996.09
2001 - 02	673.43	2971.12	1822.28	562.50	724.12	643.31	1059.25
2002 - 03	873.43	2971.12	1922.28	562.50	1433.03	997.76	1217.50
2003 - 04	873.43	2971.12	1922.28	562.50	1501.03	1031.77	1260.48
2004 - 05	873.43	2971.12	1922.28	562.50	2983.04	1772.77	1464.36
CGR	10.13	4.01	5.05	7.82	19.50	15.05	7.29
t Value	-5.522	-3.682	-4.539	-0.751	-4.146	-4.005	-1.706

Source: *Field Survey and Analyses of Sample Sugar Factories (Refer Nature and Source of Data Collection)*

APPENDIX - 11: LEVEL OF RESERVE FUNDS OF SELECTED SUGAR FACTORIES FROM 1990-91 TO 2004-05 (Rs in lakhs)

YEARS	CO-OPERATIVE SECTOR								
	ABOVE AVERAGE				*ABOVE AVERAGE*				*AVG*
	HIRANYA-KESHI	*MALA-PRABHA*	*DOOD-GHANGA*	*AVG*	*PANDAVA-PURA*	*GHATA-PRABHA*	*HEMA-VATHI*	*AVG*	
1990 - 91	57.59	3.89	2.41	21.30	0.00	1226.51	588.00	604.84	313.07
1991 - 92	65.16	3.97	3.94	24.36	44.17	882.04	659.85	528.69	276.52
1992 - 93	70.05	4.18	4.29	26.17	44.19	930.49	727.32	567.33	296.75
1993 - 94	79.70	5.78	8.62	31.37	0.00	991.99	792.48	594.82	313.10
1994 - 95	94.40	6.33	8.65	36.46	45.07	1097.87	853.41	665.45	350.96
1995 - 96	104.22	7.76	14.30	42.09	45.09	1064.83	1028.05	712.66	377.38
1996 - 97	115.06	10.76	14.53	46.78	45.10	1169.28	1190.56	801.65	424.22
1997 - 98	121.83	10.76	14.53	49.04	45.68	1232.31	1181.56	819.85	434.44
1998 - 99	126.86	16.36	15.03	52.75	45.72	1452.13	1286.08	927.98	490.36
1999 - 00	140.68	16.64	22.13	59.82	45.73	1509.48	1219.81	925.01	492.41
2000 - 01	148.20	20.31	22.63	63.71	45.74	1711.17	1222.47	993.13	528.42
2001 - 02	156.78	20.72	22.63	66.71	68.03	1978.18	1277.46	1107.89	587.30
2002 - 03	162.79	20.72	22.63	68.71	68.03	1715.25	1272.97	1018.75	543.73
2003 - 04	195.32	20.72	22.63	79.56	68.03	1926.27	1344.28	1112.86	596.21
2004 - 05	214.68	20.73	22.63	86.01	0.00	2094.42	1371.93	1155.45	620.73
CGR	8.38	12.04	11.11	8.95	4.72	5.93	5.30	5.64	5.83
t Value	-2.033	-2.204	-3.727	-3.134	-0.845	-5.068	-3.98	-3.286	-5.038

Source: *Field Survey and Analyses of Sample Sugar Factories (Refer Nature and Source of Data Collection)*

APPENDIX - 11a: LEVEL OF RESERVE FUNDS OF SELECTED SUGAR FACTORIES FROM 1990-91 TO 2004-05 (Rs in lakhs)

YEARS	*PUBLIC SECTOR*			*PRIVATE SECTOR*			*OVERALL AVG (CO-OPERATIVE, PUBLIC AND PRIVATE)*
	ABOVE AVG	*BELOW AVG*	*AVG*	*ABOVE AVG*	*BELOW AVG*	*AVG*	
	MY SUGAR	*MYSORE*		*UGAR*	*DHAVANAGERE*		
1990 - 91	605.12	1625.31	1115.21	392.08	26.58	209.33	545.87
1991 - 92	587.26	1625.31	1106.28	398.19	26.58	212.39	531.73
1992 - 93	541.34	1188.25	864.80	402.13	26.58	214.36	458.63
1993 - 94	541.44	550.26	545.85	427.93	26.58	227.26	362.07
1994 - 95	951.41	282.67	617.04	631.03	26.58	328.81	432.27
1995 - 96	1022.99	903.15	963.07	1079.24	26.58	552.91	631.12
1996 - 97	1003.81	943.69	973.75	1308.91	26.58	667.75	688.57
1997 - 98	829.25	942.41	885.83	1476.68	27.98	752.33	690.87
1998 - 99	961.63	962.81	962.22	1931.81	27.98	979.90	810.83
1999 - 00	981.97	993.51	987.74	2714.81	28.00	1371.41	950.52
2000 - 01	811.93	1101.06	956.50	4113.77	258.55	2186.16	1223.69
2001 - 02	794.13	352.21	573.17	3609.34	287.66	1948.50	1036.32
2002 - 03	784.63	202.84	493.74	3489.30	303.49	1896.40	977.95
2003 - 04	536.50	193.18	364.84	3118.20	313.50	1715.85	892.30
2004 - 05	527.01	200.90	363.95	3705.38	313.50	2009.44	998.04
CGR	0.39	-9.60	-4.73	15.26	2219.79	15.59	6.56
t Value	-0.938	4.776	4.291	-3.311	-5.042	-4.484	-4.319

Source: *Field Survey and Analyses of Sample Sugar Factories (Refer Nature and Source of Data Collection)*

APPENDIX - 12: LEVEL OF OWNED FUNDS OF SELECTED SUGAR FACTORIES FROM 1990-91 TO 2004-05 (Rs in lakhs)

YEARS	CO-OPERATIVE SECTOR								
	ABOVE AVERAGE				ABOVE AVERAGE				AVG
	HIRANYA-KESHI	MALA-PRABHA	DOOD-GHANGA	AVG	PANDAVA-PURA	GHATA-PRABHA	HEMA-VATHI	AVG	
1990 - 91	201.64	196.65	427.44	275.24	412.00	1629.45	888.00	976.48	625.86
1991 - 92	209.19	196.83	428.97	278.33	456.17	1284.98	959.85	900.33	589.33
1992 - 93	215.72	197.26	1104.29	505.76	456.19	1334.77	1027.32	939.43	722.59
1993 - 94	225.64	198.86	1108.62	511.04	412.00	1526.49	1292.48	1076.99	794.02
1994 - 95	240.67	246.45	1108.65	531.92	457.07	1607.83	1353.41	1139.44	835.68
1995 - 96	250.98	320.83	1114.30	562.04	457.09	1574.83	1528.05	1186.66	874.35
1996 - 97	327.32	323.94	1114.53	588.60	457.10	1728.30	1690.56	1291.99	940.29
1997 - 98	363.75	323.92	1212.53	633.40	457.68	1824.86	1681.56	1321.37	977.38
1998 - 99	368.93	382.24	1213.03	654.73	457.72	2050.22	1786.08	1431.34	1043.04
1999 - 00	383.00	382.51	1022.13	595.88	457.73	2107.57	1882.59	1482.63	1039.25
2000 - 01	390.76	387.53	1022.63	600.31	457.74	2309.72	1890.37	1552.61	1076.46
2001 - 02	430.91	770.98	1522.63	908.17	480.03	2642.89	1949.26	1690.73	1299.45
2002 - 03	441.96	820.97	1522.63	928.52	480.03	2380.01	2011.56	1623.87	1276.19
2003 - 04	984.60	821.38	1522.63	1109.54	480.03	2591.07	2141.90	1737.67	1423.60
2004 - 05	1007.04	821.51	1133.82	987.46	0.00	2759.26	2190.99	1650.08	1318.77
CGR	11.52	11.85	4.82	7.76	-1.97	5.28	5.78	4.71	5.71
t Value	-8.845	-6.148	0.985	-4.539	-1.055	-5.044	-4.754	-3.92	-6.532

Source: *Field Survey and Analyses of Sample Sugar Factories (Refer Nature and Source of Data Collection)*

APPENDIX - 12a: LEVEL OF OWNED FUNDS OF SELECTED SUGAR FACTORIES FROM 1990-91 TO 2004-05 (Rs in lakhs)

YEARS	PUBLIC SECTOR			PRIVATE SECTOR			OVERALL AVG (CO-OPERATIVE, PUBLIC AND PRIVATE)
	ABOVE AVG	BELOW AVG	AVG	ABOVE AVG	BELOW AVG	AVG	
	MY SUGAR	MYSORE		UGAR	DHAVANAGERE		
1990 - 91	905.60	3350.61	2128.11	588.08	228.57	408.32	1054.10
1991 - 92	860.69	3238.56	2049.62	594.19	244.51	419.35	1019.43
1992 - 93	814.77	2801.50	1808.14	602.13	262.27	432.20	987.64
1993 - 94	814.87	3013.51	1914.19	627.93	262.67	445.30	1051.17
1994 - 95	1224.84	2745.92	1985.38	994.33	265.02	629.68	1150.25
1995 - 96	1296.42	3366.41	2331.41	1454.24	265.62	859.93	1355.23
1996 - 97	1277.24	3406.95	2342.09	1683.91	265.71	974.81	1419.06
1997 - 98	1302.68	3405.66	2354.17	1926.68	394.22	1160.45	1497.34
1998 - 99	1635.06	3426.06	2530.56	2493.37	394.23	1443.80	1672.47
1999 - 00	1655.40	3930.06	2792.73	3277.31	576.68	1926.99	1919.66
2000 - 01	1485.36	4037.61	2761.48	4676.27	966.56	2821.42	2219.79
2001 - 02	1467.56	3323.33	2395.45	4171.84	1011.78	2591.81	2095.57
2002 - 03	1658.06	3173.96	2416.01	4051.80	1736.52	2894.16	2195.45
2003 - 04	1409.93	3164.30	2287.12	3680.70	1814.53	2747.61	2152.78
2004 - 05	1400.44	3172.02	2286.23	4267.88	3296.53	3782.21	2462.40
CGR	4.31	0.69	1.70	13.92	19.72	15.40	6.95
t Value	-4.851	1.56	0.069	-4.593	-0.744	-4.534	-4.51

Source: *Field Survey and Analyses of Sample Sugar Factories (Refer Nature and Source of Data Collection)*

APPENDIX - 13: CURRENT ASSETS POSITION OF SELECTED SUGAR FACTORIES FROM 1990-91 TO 2004-05 (Rs in lakhs)

YEARS	CO-OPERATIVE SECTOR								
	ABOVE AVERAGE				*ABOVE AVERAGE*				*AVG*
	HIRANYA-KESHI	*MALA-PRABHA*	*DOOD-GHANGA*	*AVG*	*PANDAVA-PURA*	*GHATA-PRABHA*	*HEMA-VATHI*	*AVG*	
1990 - 91	4467.22	3183.45	378.59	2676.42	0.00	1165.67	313.59	493.09	1584.75
1991 - 92	5077.89	3994.13	422.03	3164.68	1083.81	1304.12	868.41	1085.45	2125.07
1992 - 93	7568.56	4498.42	504.33	4190.44	1130.36	1692.69	1039.01	1287.35	2738.89
1993 - 94	7448.66	5394.91	1164.67	4669.41	0.00	1746.93	973.44	906.79	2788.10
1994 - 95	7842.49	5440.05	2659.84	5314.13	973.43	2239.58	1671.35	1628.12	3471.12
1995 - 96	9907.04	6443.02	918.62	5756.22	1771.66	2692.37	2153.44	2205.82	3981.02
1996 - 97	12437.69	7915.75	1035.19	7129.54	1279.15	3084.60	1807.57	2057.11	4593.32
1997 - 98	13422.46	5916.89	870.96	6736.77	373.49	2619.40	1153.94	1382.28	4059.52
1998 - 99	12347.78	7129.29	1185.89	6887.65	1732.45	3056.97	1779.47	2189.63	4538.64
1999 - 00	22748.45	8377.59	1501.22	10875.76	2739.69	3448.82	2281.66	2823.39	6849.57
2000 - 01	16512.61	10220.85	11957.43	12896.96	3706.48	5002.23	3245.03	3984.58	8440.77
2001 - 02	20578.13	16316.57	12312.29	16402.33	3043.96	5502.88	2724.06	3756.97	10079.65
2002 - 03	15771.87	7301.04	13068.77	12047.23	3205.97	4060.85	2745.33	3337.38	7692.30
2003 - 04	13032.29	4809.80	12548.08	10130.06	1554.79	4931.95	2659.52	3048.75	6589.41
2004 - 05	7148.09	10415.98	14539.30	10701.12	0.00	6870.20	1530.77	2800.32	6750.72
CGR	6.43	6.52	22.02	9.72	8.53	10.66	7.90	9.42	9.65
t Value	-6.266	-3.596	-4.865	-3.944	-0.155	0.549	-6.713	-4.117	-7.699

Source: *Field Survey and Analyses of Sample Sugar Factories (Refer Nature and Source of Data Collection)*

APPENDIX - 13a: CURRENT ASSETS POSITION OF SELECTED SUGAR FACTORIES FROM 1990-91 TO 2004-05 (Rs in lakhs)

YEARS	*PUBLIC SECTOR*			*PRIVATE SECTOR*			*OVERALL AVG*
	ABOVE AVG	*BELOW AVG*	*AVG*	*ABOVE AVG*	*BELOW AVG*	*AVG*	*(CO-OPERATIVE,*
	MY SUGAR	*MYSORE*		*UGAR*	*DHAVANAGERE*		*PUBLIC AND PRIVATE)*
1990 - 91	1844.48	1728.48	1786.48	5403.10	999.06	3201.08	2190.77
1991 - 92	2773.08	2335.33	2554.20	5839.40	1321.57	3580.48	2753.25
1992 - 93	3596.62	3051.33	3323.98	6658.24	1564.28	4111.26	3391.38
1993 - 94	2936.11	3178.98	3057.54	7801.11	775.60	4288.35	3378.00
1994 - 95	4413.36	4473.88	4443.62	8370.47	1817.32	5093.89	4336.21
1995 - 96	5670.86	5396.60	5533.73	10272.39	2300.31	6286.35	5267.03
1996 - 97	6315.24	4909.28	5612.26	7796.63	2377.88	5087.25	5097.61
1997 - 98	4111.64	4254.03	4182.84	6244.47	2796.64	4520.56	4254.31
1998 - 99	7017.44	4996.36	6006.90	12115.68	3404.21	7759.95	6101.83
1999 - 00	11371.25	5898.32	8634.79	16199.86	4663.56	10431.71	8638.69
2000 - 01	10836.46	5388.01	8112.23	14440.28	4392.60	9416.44	8656.48
2001 - 02	10915.08	5982.78	8448.93	13876.59	6345.56	10111.08	9546.55
2002 - 03	9059.98	525.02	4792.50	18410.00	5453.66	11931.83	8138.88
2003 - 04	9815.98	5912.70	7864.34	12358.00	6773.27	9565.63	8006.46
2004 - 05	1377.38	6095.11	3736.24	16588.43	9049.33	12818.88	7768.62
CGR	7.54	4.55	6.31	7.75	14.30	9.39	8.55
t Value	-5.199	-3.818	-1.392	-6.774	-2.934	-2.804	-3.646

Source: *Field Survey and Analyses of Sample Sugar Factories (Refer Nature and Source of Data Collection)*

APPENDIX - 14: FIXED ASSETS POSITION OF SELECTED SUGAR FACTORIES FROM 1990-91 TO 2004-05 (Rs in lakhs)

YEARS	CO-OPERATIVE SECTOR								
	ABOVE AVERAGE				ABOVE AVERAGE				AVG
	HIRANYA-KESHI	MALA-PRABHA	DOOD-GHANGA	AVG	PANDAVA-PURA	GHATA-PRABHA	HEMA-VATHI	AVG	
1990 - 91	2601.03	1276.34	2689.01	2188.79	0.00	1050.39	1466.57	838.99	1513.89
1991 - 92	3030.72	1290.15	3032.95	2451.27	768.40	1073.21	2043.77	1295.13	1873.20
1992 - 93	3434.51	1480.64	3804.68	2906.61	830.46	1163.86	2284.10	1426.14	2166.38
1993 - 94	3828.78	1539.62	4314.56	3227.65	0.00	1466.07	1956.18	1140.75	2184.20
1994 - 95	4712.96	1657.41	5823.60	4064.66	749.60	1485.50	2935.19	1723.43	2894.04
1995 - 96	5232.79	1813.84	10168.80	5738.48	927.27	1756.88	3626.40	2103.52	3921.00
1996 - 97	5735.16	2076.94	11173.40	6328.50	3282.45	1896.25	3127.97	2768.89	4548.69
1997 - 98	6365.10	2398.66	13073.67	7279.14	5929.95	1909.39	2480.23	3439.86	5359.50
1998 - 99	7219.92	2545.18	13244.67	7669.92	5569.00	1961.18	3131.17	3553.78	5611.85
1999 - 00	8067.67	2659.91	15917.20	8881.59	5576.50	1840.22	3666.46	3694.39	6287.99
2000 - 01	8932.78	2706.96	7653.87	6431.20	5581.31	2106.03	4650.49	4112.61	5271.91
2001 - 02	9454.79	2871.86	7725.77	6684.14	5589.38	2159.97	4148.07	3965.81	5324.97
2002 - 03	9698.49	2871.86	9569.33	7379.89	880.39	3140.89	4237.30	2752.86	5066.38
2003 - 04	9684.72	3026.85	13494.92	8735.50	5589.40	2292.54	4115.21	3999.05	6367.27
2004 - 05	9793.41	3343.63	11618.82	8251.95	0.00	2820.91	3093.67	1971.53	5111.74
CGR	8.98	6.78	7.51	7.96	10.07	6.56	5.43	7.35	7.77
t Value	-4.633	-2.65	-4.193	-4.99	-4.297	4.618	4.618	-0.56	-2.406

Source: *Field Survey and Analyses of Sample Sugar Factories (Refer Nature and Source of Data Collection)*

APPENDIX - 14a FIXED ASSETS POSITION OF SELECTED SUGAR FACTORIES FROM 1990-91 TO 2004-05 (Rs in lakhs)

YEARS	*PUBLIC SECTOR*			*PRIVATE SECTOR*			*OVERALL AVG*
	ABOVE AVG	*BELOW AVG*	*AVG*	*ABOVE AVG*	*BELOW AVG*	*AVG*	*(CO-OPERATIVE, PUBLIC AND PRIVATE)*
	MY SUGAR	*MYSORE*		*UGAR*	*DHAVANAGERE*		
1990 - 91	1882.73	8915.47	5399.10	2306.19	120.32	1213.25	2708.75
1991 - 92	1995.52	7251.64	4623.58	2541.88	150.23	1346.05	2614.28
1992 - 93	2047.50	6270.46	4158.98	861.37	126.29	493.83	2273.06
1993 - 94	2363.15	4765.74	3564.45	3203.00	119.31	1661.15	2469.93
1994 - 95	2640.56	3367.62	3004.09	1925.90	142.64	1034.27	2310.80
1995 - 96	2579.20	3855.80	3217.50	2867.51	226.20	1546.85	2895.12
1996 - 97	2631.96	3544.52	3088.24	3546.30	209.43	1877.87	3171.60
1997 - 98	2829.02	3231.69	3030.35	7067.45	222.47	3644.96	4011.60
1998 - 99	3004.47	3307.16	3155.82	7437.97	951.67	4194.82	4320.83
1999 - 00	3652.82	3226.50	3439.66	6714.54	2367.59	4541.06	4756.24
2000 - 01	4296.84	7354.01	5825.43	9290.71	2891.16	6090.93	5729.42
2001 - 02	5175.79	6963.77	6069.78	9304.19	2774.52	6039.35	5811.37
2002 - 03	7335.22	6397.67	6866.44	9300.50	2757.21	6028.85	5987.22
2003 - 04	78535.48	6841.86	42688.67	10831.24	8407.93	9619.59	19558.51
2004 - 05	9608.83	7374.37	8491.60	9353.59	8224.97	8789.28	7464.21
CGR	22.91	0.43	14.20	12.34	25.86	15.79	12.82
t value	0.273	-1.752	0.403	-4.003	-1.075	-2.872	0.283

Source: *Field Survey and Analyses of Sample Sugar Factories (Refer Nature and Source of Data Collection)*

APPENDIX - 15: TOTAL ASSETS POSITION OF SELECTED SUGAR FACTORIES FROM 1990-91 TO 2004-05 (Rs in lakhs)

YEARS	CO-OPERATIVE SECTOR								
	ABOVE AVERAGE				ABOVE AVERAGE				AVG
	HIRANYA-KESHI	MALA-PRABHA	DOOD-GHANGA	AVG	PANDAVA-PURA	GHATA-PRABHA	HEMA-VATHI	AVG	
1990 - 91	7319.59	4459.79	3067.59	4948.99	0.00	2216.06	251.86	822.64	2885.82
1991 - 92	8423.53	5284.28	3454.99	5720.93	2164.88	2377.33	103.61	1548.61	3634.77
1992 - 93	11374.49	5979.06	4309.01	7220.85	2989.87	2856.55	560.00	2135.47	4678.16
1993 - 94	11797.42	6934.53	5479.23	8070.39	0.00	3213.00	344.49	1185.83	4628.11
1994 - 95	13029.99	7097.46	8483.44	9536.96	2485.47	3725.08	279.22	2163.26	5850.11
1995 - 96	15523.44	8256.86	11087.42	11622.57	3355.40	4449.25	1306.10	3036.92	7329.74
1996 - 97	18519.84	9992.69	12208.58	13573.70	5768.31	4980.85	1824.98	4191.38	8882.54
1997 - 98	20122.09	8315.55	13944.63	14127.42	7329.88	4528.79	1448.97	4435.88	9281.65
1998 - 99	19920.83	9674.47	14430.55	14675.28	11425.53	5018.15	427.91	5623.86	10149.57
1999 - 00	23410.33	11037.50	17418.42	17288.75	14403.15	5289.04	1392.49	7028.23	12158.49
2000 - 01	25780.14	12927.81	19611.30	19439.75	17183.02	7108.26	2038.51	8776.60	14108.17
2001 - 02	30339.63	19188.43	20038.06	23188.71	18053.73	7662.85	2712.86	9476.48	16332.59
2002 - 03	25689.76	10172.90	22638.09	19500.25	20427.39	7201.74	2378.05	10002.39	14751.32
2003 - 04	23692.45	7836.65	26043.00	19190.70	19757.29	7224.49	2494.80	9825.53	14508.11
2004 - 05	16941.50	13759.61	26158.12	18953.08	0.00	9691.11	2397.96	4029.69	11491.38
CGR	7.10	6.58	12.73	8.87	14.39	9.17	13.95	13.95	9.84
t Value	-2.977	-10.028	-2.992	-3.833	-4.42	-3.418	-0.486	0.891	-0.674

Source: *Field Survey and Analyses of Sample Sugar Factories (Refer Nature and Source of Data Collection)*

APPENDIX - 15a TOTAL ASSETS POSITION OF SELECTED SUGAR FACTORIES FROM 1990-91 TO 2004-05 (Rs in lakhs)

YEARS	*PUBLIC SECTOR*			*PRIVATE SECTOR*			*OVERALL AVG*
	ABOVE AVG	*BELOW AVG*	*AVG*	*ABOVE AVG*	*BELOW AVG*	*AVG*	*(CO-OPERATIVE, PUBLIC AND PRIVATE)*
	MY SUGAR	*MYSORE*		*UGAR*	*DHAVANAGERE*		
1990 - 91	1507.47	613602.50	307554.99	7709.29	1119.38	4414.33	104951.71
1991 - 92	1836.06	5747.18	3791.62	8381.28	1471.80	4926.54	4117.64
1992 - 93	2537.77	5112.24	3825.00	7519.61	1690.57	4605.09	4369.42
1993 - 94	1773.57	5716.14	3744.86	11004.11	894.91	5949.51	4774.16
1994 - 95	2613.17	5654.22	4133.70	10296.37	1959.96	6128.16	5370.66
1995 - 96	4249.67	6225.75	5237.71	13139.90	2526.51	7833.20	6800.22
1996 - 97	5686.50	6405.82	6046.16	11342.93	2587.31	6965.12	7297.94
1997 - 98	3743.31	7388.07	5565.69	13311.92	3019.11	8165.52	7670.95
1998 - 99	5227.23	7751.04	6489.14	19553.65	4355.88	11954.77	9531.16
1999 - 00	7097.65	8791.67	7944.66	22914.40	7031.14	14972.77	11691.97
2000 - 01	7473.91	9126.83	8300.37	23730.99	7283.75	15507.37	12638.64
2001 - 02	8959.63	7566.76	8263.20	23180.78	9120.08	16150.43	13582.07
2002 - 03	11138.43	7472.18	9305.31	27710.50	8210.87	17960.68	14005.77
2003 - 04	1274.99	7565.35	4420.17	23189.24	15181.20	19185.22	12704.50
2004 - 05	727.39	7358.40	4042.89	25942.02	17274.30	21608.16	12380.81
CGR	6.73	-31.60	-28.35	9.34	18.40	11.62	-11.07
t Value	-3.812	0.981	0.825	-5.892	-1.288	-4.471	0.225

Source: *Field Survey and Analyses of Sample Sugar Factories (Refer Nature and Source of Data Collection)*

APPENDIX - 16: CURRENT LIABILITIES OF SELECTED SUGAR FACTORIES FROM 1990-91 TO 2004-05 (Rs in lakhs)

YEARS	CO-OPERATIVE SECTOR								
	ABOVE AVERAGE				*ABOVE AVERAGE*				*AVG*
	HIRANYA-KESHI	*MALA-PRABHA*	*DOOD-GHANGA*	*AVG*	*PANDAVA-PURA*	*GHATA-PRABHA*	*HEMA-VATHI*	*AVG*	
1990 - 91	1975.64	1911.24	66.53	1317.80	0.00	372.23	406.94	259.72	788.76
1991 - 92	2447.13	2358.08	85.48	1630.23	233.41	411.83	684.29	443.18	1036.70
1992 - 93	4310.61	2804.96	52.36	2389.31	213.67	513.52	427.71	384.97	1387.14
1993 - 94	4273.45	4430.63	450.08	3051.39	0.00	630.04	467.80	365.95	1708.67
1994 - 95	3320.53	3341.44	451.55	2371.17	213.95	792.37	855.81	620.71	1495.94
1995 - 96	3504.52	3671.90	897.81	2691.41	244.71	1116.34	969.37	776.81	1734.11
1996 - 97	5049.35	3753.41	1393.33	3398.70	251.07	1106.03	605.24	654.11	2026.40
1997 - 98	5123.37	3558.52	1788.03	3489.97	795.97	307.60	395.12	499.56	1994.77
1998 - 99	3796.90	4260.46	2253.49	3436.95	980.01	1130.00	1304.96	1138.32	2287.64
1999 - 00	3915.11	4084.94	2714.82	3571.62	1307.06	1063.57	1048.51	1139.71	2355.67
2000 - 01	4172.98	4524.31	2658.58	3785.29	1830.69	1680.47	1593.57	1701.58	2743.43
2001 - 02	4364.24	3208.26	2106.18	3226.23	822.99	1250.98	1214.02	1096.00	2161.11
2002 - 03	3709.44	3252.90	1616.57	2859.64	1927.00	1617.82	843.89	1462.90	2161.27
2003 - 04	1429.77	2812.28	1388.80	1876.95	805.71	740.26	1130.77	892.25	1384.60
2004 - 05	3902.95	1067.21	1518.34	2162.83	0.00	1320.34	944.77	755.04	1458.94
CGR	0.67	-0.10	12.00	2.14	12.80	7.32	6.06	8.32	3.55
t Value	-2.521	0.066	-1.342	-5.234	-1.092	-6.558	-3.779	-2.595	-2.766

Source: *Field Survey and Analyses of Sample Sugar Factories (Refer Nature and Source of Data Collection)*

APPENDIX - 16a CURRENT LIABILITIES OF SELECTED SUGAR FACTORIES FROM 1990-91 TO 2004-05 (Rs in lakhs)

YEARS	PUBLIC SECTOR			PRIVATE SECTOR			OVERALL AVG
	ABOVE AVG	BELOW AVG	AVG	ABOVE AVG	BELOW AVG	AVG	(CO-OPERATIVE, PUBLIC AND PRIVATE)
	MY SUGAR	MYSORE		UGAR	DHAVANAGERE		
1990 - 91	1263.48	581.16	922.32	2862.10	766.49	1814.30	1175.13
1991 - 92	1518.90	785.81	1152.35	3308.06	1133.31	2220.68	1469.91
1992 - 93	1880.55	0.00	940.28	3900.08	1298.61	2599.34	1642.25
1993 - 94	2049.27	3178.98	2614.12	4591.44	484.12	2537.78	2286.86
1994 - 95	2751.64	1616.98	2184.31	4367.84	496.63	2432.23	2037.49
1995 - 96	2550.92	2192.32	2371.62	4586.06	806.47	2696.27	2267.33
1996 - 97	1758.95	1610.79	1684.87	4749.71	512.07	2630.89	2114.05
1997 - 98	1635.75	1933.85	1784.80	4397.56	709.93	2553.74	2111.10
1998 - 99	3161.52	2677.40	2919.46	4012.14	1179.20	2595.67	2600.92
1999 - 00	6194.63	3694.76	4944.69	6207.62	1844.45	4026.04	3775.47
2000 - 01	5799.72	3173.95	4486.83	5677.46	1519.38	3598.42	3609.56
2001 - 02	5183.86	4300.36	4742.11	4112.33	1922.28	3017.30	3306.84
2002 - 03	5885.59	3872.26	4878.92	6297.71	1395.58	3846.65	3628.95
2003 - 04	6540.83	4023.41	5282.12	6821.09	3235.75	5028.42	3898.38
2004 - 05	11331.43	4384.61	7858.02	6583.01	3244.69	4913.85	4743.60
CGR	13.59	11.08	12.61	4.70	10.95	6.08	8.14
t Value	-1.68	0.935	-3.905	-4.685	-0.112	0.335	-3.946

Source: *Field Survey and Analyses of Sample Sugar Factories (Refer Nature and Source of Data Collection)*

APPENDIX - 17: TOTAL BORROWINGS OF SELECTED SUGAR FACTORIES FROM 1990-91 TO 2004-05 (Rs in lakhs)

YEARS	CO-OPERATIVE SECTOR								
	ABOVE AVERAGE				ABOVE AVERAGE				AVG
	HIRANYA-KESHI	MALA-PRABHA	DOOD-GHANGA	AVG	PANDAVA-PURA	GHATA-PRABHA	HEMA-VATHI	AVG	
1990 - 91	1352.73	793.70	551.30	899.24	0.00	1226.51	203.58	476.70	687.97
1991 - 92	1614.54	1910.90	1000.67	1508.70	780.37	1194.00	313.44	762.60	1135.65
1992 - 93	2226.36	1123.60	1187.80	1512.59	1023.68	1251.98	518.97	931.54	1222.06
1993 - 94	2536.69	726.16	1459.77	1574.21	0.00	1245.77	529.09	591.62	1082.91
1994 - 95	4475.08	1864.83	3702.14	3347.35	723.21	2001.91	743.56	1156.23	2251.79
1995 - 96	6489.49	2523.44	5443.40	4818.78	1232.51	2048.88	1020.00	1433.80	3126.29
1996 - 97	6757.91	4012.74	5562.78	5444.48	3657.87	2262.12	1081.29	2333.76	3889.12
1997 - 98	6596.64	2345.57	5087.06	4676.42	4850.78	2245.64	675.18	2590.53	3633.48
1998 - 99	6402.11	2804.03	4515.56	4573.90	6958.75	2485.78	608.24	3350.92	3962.41
1999 - 00	8726.64	4268.18	5502.78	6165.87	8817.91	2339.80	1417.24	4191.65	5178.76
2000 - 01	10050.35	5526.65	6218.86	7265.29	9670.03	2873.17	1089.82	4544.34	5904.81
2001 - 02	13947.60	6586.89	6984.46	9172.98	10347.76	3789.41	1786.24	5307.80	7240.39
2002 - 03	10719.34	4123.35	7579.65	7474.11	10401.74	2849.59	1522.79	4924.71	6199.41
2003 - 04	12516.67	2816.62	11449.79	8927.69	10286.26	3454.92	1824.73	5188.64	7058.16
2004 - 05	17807.60	2859.63	11942.25	10869.83	0.00	3413.88	539.00	1317.63	6093.73
CGR	13.82	8.66	13.92	12.88	14.85	7.79	8.81	12.05	12.60
t Value	0.862	-2.687	-2.656	-4.624	-5.119	-2.438	-4.461	-5.76	-3.374

Source: *Field Survey and Analyses of Sample Sugar Factories (Refer Nature and Source of Data Collection)*

APPENDIX - 17a TOTAL BORROWINGS OF SELECTED SUGAR FACTORIES FROM 1990-91 TO 2004-05 (Rs in lakhs)

YEARS	*PUBLIC SECTOR*			*PRIVATE SECTOR*			*OVERALL AVG*
	ABOVE AVG	*BELOW AVG*	*AVG*	*ABOVE AVG*	*BELOW AVG*	*AVG*	*(CO-OPERATIVE, PUBLIC AND PRIVATE)*
	MY SUGAR	*MYSORE*		*UGAR*	*DHAVANAGERE*		
1990 - 91	601.87	2489.52	1545.70	3058.12	517.27	1787.70	1340.45
1991 - 92	975.37	2489.52	1732.45	3159.92	458.91	1809.42	1559.17
1992 - 93	1723.01	2291.14	2007.07	3210.08	439.00	1824.54	1684.56
1993 - 94	948.61	2682.61	1815.61	3784.02	385.68	2084.85	1661.12
1994 - 95	1384.09	2888.20	2136.15	5303.41	1374.36	3338.88	2575.61
1995 - 96	2953.25	2837.23	2895.24	7505.41	1636.23	4570.82	3530.78
1996 - 97	4409.26	2974.39	3691.83	5361.14	1987.14	3674.14	3751.70
1997 - 98	2440.63	3957.91	3199.27	7492.15	2050.87	4771.51	3868.09
1998 - 99	3592.17	3811.64	3701.91	13080.91	2897.89	7989.40	5217.91
1999 - 00	5442.25	4815.35	5128.80	13463.64	4676.11	9069.87	6459.14
2000 - 01	5988.55	5030.57	5509.56	13400.77	4808.16	9104.47	6839.61
2001 - 02	7424.59	4218.78	5821.68	14513.12	6191.38	10352.25	7804.78
2002 - 03	9480.36	4234.39	6857.38	16707.23	5084.69	10895.96	7984.25
2003 - 04	11569.80	4349.15	7959.47	17819.10	10450.57	14134.84	9717.49
2004 - 05	12137.43	3713.28	7925.35	15583.29	12422.58	14002.93	9340.67
CGR	17.11	4.75	11.84	12.32	20.03	14.47	13.23
t Value	-2.692	-6.472	-1.717	-3.485	1.885	-3.259	-6.348

Source: *Field Survey and Analyses of Sample Sugar Factories (Refer Nature and Source of Data Collection)*

APPENDIX - 18: TOTAL LIABILITIES OF SELECTED SUGAR FACTORIES FROM 1990-91 TO 2004-05 (Rs in lakhs)

YEARS	CO-OPERATIVE SECTOR								
	ABOVE AVERAGE				ABOVE AVERAGE				AVG
	HIRANYA-KESHI	MALA-PRABHA	DOOD-GHANGA	AVG	PANDAVA-PURA	GHATA-PRABHA	HEMA-VATHI	AVG	
1990 - 91	3328.37	2704.94	617.82	2217.04	0.00	1598.74	610.52	736.42	1476.73
1991 - 92	4061.67	4268.98	1086.14	3138.93	1013.79	1605.83	997.73	1205.78	2172.36
1992 - 93	6536.97	3928.56	1240.16	3901.90	1237.35	1765.50	946.68	1316.51	2609.20
1993 - 94	6810.14	5156.79	1909.85	4625.59	0.00	1875.81	996.89	957.57	2791.58
1994 - 95	7795.61	5206.27	4153.69	5718.52	937.16	2794.28	1599.37	1776.94	3747.73
1995 - 96	9994.01	6195.34	6341.21	7510.19	1477.22	3165.22	1989.37	2210.60	4860.40
1996 - 97	11807.26	7766.15	6956.11	8843.17	3908.94	3368.15	1686.53	2987.87	5915.52
1997 - 98	11720.01	5904.09	6875.09	8166.40	5646.75	2553.24	1070.30	3090.10	5628.25
1998 - 99	10199.01	7064.49	6769.05	8010.85	7938.76	3615.78	1913.20	4489.25	6250.05
1999 - 00	12641.75	8353.12	8217.60	9737.49	10124.97	3403.37	2465.75	5331.36	7534.43
2000 - 01	14223.33	10050.96	8877.44	11050.58	11500.72	4553.64	2683.39	6245.92	8648.25
2001 - 02	18311.84	9795.15	9090.64	12399.21	11170.75	5040.39	3000.26	6403.80	9401.50
2002 - 03	14428.78	7376.25	9196.22	10333.75	12328.74	4467.41	2366.68	6387.61	8360.68
2003 - 04	13946.44	5628.90	12838.60	10804.65	11091.97	4195.18	2955.50	6080.88	8442.76
2004 - 05	21710.55	3926.84	13460.60	13032.66	0.00	4734.22	1483.77	2072.66	7552.66
CGR	9.48	4.06	13.53	9.17	14.60	7.66	7.49	11.16	9.77
t Value	-5.465	-2.76	-2.264	-5.051	-4.605	-4.24	-6.05	-5.277	-4.35

Source: *Field Survey and Analyses of Sample Sugar Factories (Refer Nature and Source of Data Collection)*

APPENDIX - 18a TOTAL LIABILITIES OF SELECTED SUGAR FACTORIES FROM 1990-91 TO 2004-05 (Rs in lakhs)

YEARS	PUBLIC SECTOR			PRIVATE SECTOR			OVERALL AVG (CO-OPERATIVE, PUBLIC AND PRIVATE)
	ABOVE AVG MY SUGAR	BELOW AVG MYSORE	AVG	ABOVE AVG UGAR	BELOW AVG DHAVANAGERE	AVG	
1990 - 91	1865.35	3070.68	2468.01	5920.22	1283.77	3601.99	2515.58
1991 - 92	2494.27	3275.33	2884.80	6467.98	1592.22	4030.10	3029.08
1992 - 93	3603.56	2291.14	2947.35	7110.16	1737.61	4423.89	3326.81
1993 - 94	2997.88	5861.59	4429.74	8375.46	869.79	4622.63	3947.98
1994 - 95	4135.73	4505.18	4320.45	9671.25	1870.98	5771.12	4613.10
1995 - 96	5504.17	5029.55	5266.86	12091.47	2442.70	7267.08	5798.11
1996 - 97	6168.21	4585.18	5376.70	10110.85	2499.21	6305.03	5865.75
1997 - 98	4076.38	5891.76	4984.07	11889.71	2760.80	7325.25	5979.19
1998 - 99	6753.69	6489.04	6621.36	17093.05	4077.09	10585.07	7818.83
1999 - 00	11636.88	8510.11	10073.49	19671.26	6520.56	13095.91	10234.61
2000 - 01	11788.27	8204.51	9996.39	19078.23	6327.55	12702.89	10449.18
2001 - 02	12608.45	8519.14	10563.79	18625.45	8113.66	13369.56	11111.62
2002 - 03	15365.95	8106.65	11736.30	23004.94	6480.28	14742.61	11613.20
2003 - 04	18110.63	8372.55	13241.59	24640.19	13686.32	19163.26	13615.87
2004 - 05	23468.86	8097.89	15783.37	22166.30	15667.27	18916.78	14084.27
CGR	15.51	7.40	12.18	9.77	17.57	11.80	11.42
t Value	-5.607	-3.803	2.218	-5.134	-1.214	-3.651	-5.905

Source: *Field Survey and Analyses of Sample Sugar Factories (Refer Nature and Source of Data Collection)*

APPENDIX - 19: TOTAL SALES OF SELECTED SUGAR FACTORIES FROM 1990-91 TO 2004-05 (Rs in lakhs)

YEARS	CO-OPERATIVE SECTOR								
	ABOVE AVERAGE				ABOVE AVERAGE				AVG
	HIRANYA-KESHI	MALA-PRABHA	DOOD-GHANGA	AVG	PANDAVA-PURA	GHATA-PRABHA	HEMA-VATHI	AVG	
1990 - 91	4995.60	3309.33	2764.11	3689.68	1035.87	1370.56	1104.99	1170.47	2430.08
1991 - 92	5585.39	4005.02	2951.36	4180.59	1710.36	2116.52	1405.62	1744.17	2962.38
1992 - 93	6484.78	5154.83	3250.59	4963.40	0.00	2088.93	1209.10	1099.34	3031.37
1993 - 94	10940.06	7068.28	4100.20	7369.51	2096.83	2734.24	2042.17	2291.08	4830.30
1994 - 95	7967.43	6643.87	3887.60	6166.30	259.06	2592.31	1939.91	1597.09	3881.70
1995 - 96	10805.00	6958.47	4321.50	7361.66	2361.09	3690.95	2003.07	2685.04	5023.35
1996 - 97	10959.69	7874.30	6075.21	8303.06	1078.17	4051.19	2485.13	2538.16	5420.61
1997 - 98	13614.93	8692.50	6447.14	9584.86	1048.55	4241.31	2724.22	2671.36	6128.11
1998 - 99	16029.86	6305.03	8174.82	10169.90	2924.66	3559.77	1929.32	2804.58	6487.24
1999 - 00	15015.19	7942.28	7948.18	10301.88	2891.66	4077.26	2342.26	3103.73	6702.81
2000 - 01	16254.55	6395.81	9131.61	10593.99	2580.57	5229.25	2515.48	3441.77	7017.88
2001 - 02	13985.31	6732.39	9560.29	10092.66	1166.71	5528.66	2781.32	3158.90	6625.78
2002 - 03	17484.99	8212.90	8575.85	11424.58	1826.84	6953.42	2508.44	3762.90	7593.74
2003 - 04	12203.80	5153.07	7309.59	8222.16	1035.87	3671.12	2296.89	2334.63	5278.39
2004 - 05	15448.42	3876.90	11431.34	10252.22	0.00	5941.46	2930.98	2957.48	6604.85
CGR	6.58	1.39	8.96	5.87	0.97	7.87	7.87	5.66	5.82
t Value	-4.187	-4.489	-4.477	-5.054	1.195	-5.167	-2.531	-1.53	-4.016

Source: *Field Survey and Analyses of Sample Sugar Factories (Refer Nature and Source of Data Collection)*

APPENDIX - 19a TOTAL SALES OF SELECTED SUGAR FACTORIES FROM 1990-91 TO 2004-05 (Rs in lakhs)

YEARS	PUBLIC SECTOR			PRIVATE SECTOR			OVERALL AVG
	ABOVE AVG	BELOW AVG	AVG	ABOVE AVG	BELOW AVG	AVG	(CO-OPERATIVE, PUBLIC AND PRIVATE)
	MY SUGAR	MYSORE		UGAR	DHAVANAGERE		
1990 - 91	4014.72	1667.31	2841.02	7386.42	676.00	4031.21	3100.77
1991 - 92	4308.81	2288.11	3298.46	7439.94	1348.27	4394.11	3551.65
1992 - 93	4109.96	2772.26	3441.11	8014.26	1355.51	4684.88	3719.12
1993 - 94	8046.15	3970.55	6008.35	9326.73	2426.09	5876.41	5571.68
1994 - 95	11763.27	2831.19	7297.23	16761.49	269.11	8515.30	6564.74
1995 - 96	11723.08	4151.22	7937.15	15181.62	2230.93	8706.27	7222.26
1996 - 97	11123.88	4881.55	8002.72	16752.74	2631.27	9692.00	7705.11
1997 - 98	12307.18	6193.67	9250.43	17856.12	3390.13	10623.13	8667.22
1998 - 99	13534.54	4326.34	8930.44	13900.67	2787.08	8343.88	7920.52
1999 - 00	13800.98	5197.79	9499.39	20737.49	2816.75	11777.12	9326.44
2000 - 01	17705.26	5482.01	11593.64	19726.39	5394.74	12560.56	10390.69
2001 - 02	16979.68	7106.00	12042.84	20057.98	4087.61	12072.80	10247.14
2002 - 03	19089.80	8491.75	13790.78	17145.92	8355.00	12750.46	11378.32
2003 - 04	13112.39	4179.36	8645.88	19430.72	4951.06	12190.89	8705.05
2004 - 05	4896.64	4179.36	4538.00	20628.41	3153.82	11891.12	7677.99
CGR	6.24	6.29	6.26	6.29	11.36	7.13	6.51
t Value	-4.524	-5.506	-4.957	-3.172	1.193	-3.51	5.915

Source: *Field Survey and Analyses of Sample Sugar Factories (Refer Nature and Source of Data Collection)*

APPENDIX - 20: GROSS INCOME OF SELECTED SUGAR FACTORIES FROM 1990-91 TO 2004-05 (Rs in lakhs)

YEARS	*CO-OPERATIVE SECTOR*								
	ABOVE AVERAGE				*ABOVE AVERAGE*				*AVG*
	HIRANYA-KESHI	*MALA-PRABHA*	*DOOD-GHANGA*	*AVG*	*PANDAVA-PURA*	*GHATA-PRABHA*	*HEMA-VATHI*	*AVG*	
1990 - 91	8619.24	430.83	4353.74	4467.94	79.27	211.32	270.58	187.06	2327.50
1991 - 92	10294.48	541.93	4893.87	5243.43	270.82	449.29	376.53	365.55	2804.49
1992 - 93	13010.90	545.81	5837.04	6464.58	0.00	417.03	277.73	231.59	3348.08
1993 - 94	16844.80	583.83	7247.67	8225.43	87.71	425.78	177.43	230.31	4227.87
1994 - 95	14539.23	634.91	7229.04	7467.73	213.61	281.65	390.42	295.23	3881.48
1995 - 96	19401.18	981.63	9169.82	9850.88	146.68	661.51	365.23	391.14	5121.01
1996 - 97	22066.22	898.73	11036.14	11333.70	320.31	892.55	379.52	530.79	5932.24
1997 - 98	25951.37	972.11	12826.43	13249.97	30.37	899.21	3026.74	1318.77	7284.37
1998 - 99	26908.47	741.10	13868.82	13839.46	21.95	292.66	392.15	235.59	7037.53
1999 - 00	28076.70	783.89	16347.12	15069.24	20.71	1006.54	452.60	493.28	7781.26
2000 - 01	30099.43	454.08	20241.91	16931.81	15.80	1315.07	277.09	535.99	8733.90
2001 - 02	33250.50	488.46	19881.57	17873.51	1753.63	948.97	437.20	1046.60	9460.06
2002 - 03	32561.54	-647.05	18654.21	16856.23	215.14	544.78	23.49	261.14	8558.68
2003 - 04	24038.38	-578.64	16763.62	13407.79	79.27	743.55	315.88	379.57	6893.68
2004 - 05	1519.77	-689.83	11431.34	4087.09	0.00	1456.99	632.45	696.48	2391.79
CGR	4.62	-19.20	8.69	5.81	8.05	8.33	1.23	5.73	5.80
t Value	-5.224	-2.25	-5.573	-4.794	1.738	-2.513	-0.618	-2.597	-1.861

Source: *Field Survey and Analyses of Sample Sugar Factories (Refer Nature and Source of Data Collection)*

APPENDIX - 20a GROSS INCOME OF SELECTED SUGAR FACTORIES FROM 1990-91 TO 2004-05 (Rs in lakhs)

YEARS	*PUBLIC SECTOR*			*PRIVATE SECTOR*			*OVERALL AVG (CO-OPERATIVE, PUBLIC AND PRIVATE)*
	ABOVE AVG	*BELOW AVG*	*AVG*	*ABOVE AVG*	*BELOW AVG*	*AVG*	
	MY SUGAR	*MYSORE*		*UGAR*	*DHAVANAGERE*		
1990 - 91	4666.93	4320.94	4493.94	7326.03	702.53	4014.28	3611.90
1991 - 92	5039.16	5290.12	5164.64	8951.83	1381.08	5166.46	4378.53
1992 - 93	5110.94	5206.31	5158.62	10232.46	1371.62	5802.04	4769.58
1993 - 94	7189.11	6249.93	6719.52	11978.28	2443.54	7210.91	6052.77
1994 - 95	13334.25	7603.64	10468.95	19870.41	1712.84	10791.62	8380.68
1995 - 96	13506.94	10041.01	11773.98	18768.07	2251.75	10509.91	9134.96
1996 - 97	11748.45	7115.41	9431.93	16651.47	2640.86	9646.16	8336.78
1997 - 98	10279.64	7541.34	8910.49	20254.77	3398.60	11826.68	9340.52
1998 - 99	16034.07	7927.82	11980.95	22895.62	2804.60	12850.11	10622.86
1999 - 00	18025.12	9254.26	13639.69	26681.52	2843.84	14762.68	12061.21
2000 - 01	17892.53	103619.90	60756.22	25464.73	5461.75	15463.24	28317.78
2001 - 02	16750.75	7935.93	12343.34	23511.06	4119.75	13815.40	11872.93
2002 - 03	14836.66	7977.66	11407.16	24786.89	8407.74	16597.32	12187.72
2003 - 04	10205.05	8742.72	9473.88	25530.81	4944.51	15237.66	10535.07
2004 - 05	4558.36	810.96	2684.66	27441.93	6175.04	16808.49	7294.98
CGR	4.56	8.16	6.51	7.34	12.07	8.05	6.97
t Value	0.833	0.479	-0.378	-3.079	-3.845	-3.729	-1.504

Source: *Field Survey and Analyses of Sample Sugar Factories (Refer Nature and Source of Data Collection)*

APPENDIX - 21: TOTAL EXPENSES OF SELECTED SUGAR FACTORIES FROM 1990-91 TO 2004-05 (Rs in lakhs)

YEARS	CO-OPERATIVE SECTOR								
	ABOVE AVERAGE				*ABOVE AVERAGE*				*AVG*
	HIRANYA-KESHI	*MALA-PRABHA*	*DOOD-GHANGA*	*AVG*	*PANDAVA-PURA*	*GHATA-PRABHA*	*HEMA-VATHI*	*AVG*	
1990 - 91	8588.97	1001.45	2828.44	4139.62	191.04	263.22	314.68	256.32	2197.97
1991 - 92	10274.91	1168.34	2884.76	4776.00	286.26	299.03	293.69	292.99	2534.50
1992 - 93	12992.88	1349.16	3435.51	5925.85	0.00	661.28	279.94	313.74	3119.80
1993 - 94	16825.53	1826.18	4196.52	7616.08	196.64	609.87	131.92	312.81	3964.44
1994 - 95	14527.79	1584.15	3887.11	6666.35	251.94	767.31	397.51	472.25	3569.30
1995 - 96	19562.08	1874.38	5255.12	8897.19	332.43	1120.68	346.34	599.82	4748.51
1996 - 97	22054.05	2066.26	5638.89	9919.73	541.81	1127.81	326.23	665.28	5292.51
1997 - 98	25933.99	1786.59	6951.99	11557.52	2729.31	924.84	159.57	1271.24	6414.38
1998 - 99	26888.02	2244.38	7405.00	12179.13	1859.99	937.91	425.89	1074.60	6626.87
1999 - 00	28056.22	3230.27	9336.83	13541.11	1646.77	1099.30	434.07	1060.05	7300.58
2000 - 01	30074.95	2897.06	9726.69	14232.90	2057.99	1736.63	596.65	1463.76	7848.33
2001 - 02	33229.40	2760.35	7708.64	14566.13	1936.09	1824.47	479.14	1413.23	7989.68
2002 - 03	32538.39	2710.20	7729.07	14325.89	1445.89	1592.45	553.67	1197.34	7761.61
2003 - 04	24823.45	1836.99	5038.14	10566.19	191.04	1575.66	491.01	752.57	5659.38
2004 - 05	6915.90	1001.60	2016.32	3311.27	0.00	1760.19	431.24	730.48	2020.88
CGR	5.26	3.90	4.47	5.01	8.48	9.97	5.33	8.66	5.29
t Value	-2.644	-5.217	-6.561	-4.702	-4.817	-4.847	-6.444	-2.441	-2.308

Source: *Field Survey and Analyses of Sample Sugar Factories (Refer Nature and Source of Data Collection)*

APPENDIX - 21a TOTAL EXPENSES OF SELECTED SUGAR FACTORIES FROM 1990-91 TO 2004-05 (Rs in lakhs)

YEARS	*PUBLIC SECTOR*			*PRIVATE SECTOR*			*OVERALL AVG*
	ABOVE AVG	*BELOW AVG*	*AVG*	*ABOVE AVG*	*BELOW AVG*	*AVG*	*(CO-OPERATIVE, PUBLIC AND PRIVATE)*
	MY SUGAR	*MYSORE*		*UGAR*	*DHAVANAGERE*		
1990 - 91	4369.67	3799.41	4084.54	7951.52	667.74	4309.63	3530.71
1991 - 92	5065.34	4838.07	4951.70	8564.82	1353.05	4958.94	4148.38
1992 - 93	5336.36	5228.15	5282.25	9871.56	1316.00	5593.78	4665.28
1993 - 94	6967.08	6095.57	6531.32	11823.43	2371.77	7097.60	5864.46
1994 - 95	12975.03	6967.35	9971.19	19813.43	1669.33	10741.38	8093.96
1995 - 96	13431.05	8560.86	10995.95	18475.39	2239.50	10357.45	8700.63
1996 - 97	12300.89	7057.67	9679.28	16361.60	2635.92	9498.76	8156.85
1997 - 98	10452.21	7533.54	8992.88	19529.91	444.17	9987.04	8464.77
1998 - 99	15885.09	7898.46	11891.78	22387.40	2765.99	12576.69	10365.11
1999 - 00	17991.05	9212.57	13601.81	25655.04	2806.16	14230.60	11710.99
2000 - 01	18045.85	10151.76	14098.81	23765.25	5377.26	14571.26	12172.80
2001 - 02	16818.25	8829.11	12823.68	23613.89	4113.10	13863.50	11558.95
2002 - 03	17425.04	8465.63	12945.33	24625.53	8392.36	16508.95	12405.30
2003 - 04	11458.71	8499.06	9978.88	23941.25	5308.79	14625.02	10087.76
2004 - 05	6568.59	8859.62	7714.11	26483.25	7232.89	16858.07	8864.35
CGR	5.55	4.60	5.18	7.11	13.69	8.07	6.44
t Value	-3.263	-5.79	-2.754	-4.071	0.718	-2.967	-3.932

Source: *Field Survey and Analyses of Sample Sugar Factories (Refer Nature and Source of Data Collection)*

APPENDIX - 22: NET PROFIT (+) / LOSS (-) OF SELECTED SUGAR FACTORIES FROM 1990-91 TO 2004-05 (Rs in lakhs)

YEARS	CO-OPERATIVE SECTOR								
	ABOVE AVERAGE				*ABOVE AVERAGE*				*AVG*
	HIRANYA-KESHI	MALA-PRABHA	DOOD-GHANGA	*AVG*	PANDAVA-PURA	GHATA-PRABHA	HEMA-VATHI	*AVG*	
1990 - 91	30.28	1.17	14.22	15.22	0.00	47.03	270.57	105.87	60.54
1991 - 92	19.56	2.31	3.11	8.33	-91.71	157.96	196.18	87.48	47.90
1992 - 93	18.01	3.58	13.66	11.75	-220.62	-474.78	78.61	-205.60	-96.92
1993 - 94	19.26	1.64	8.68	9.86	-278.94	388.88	8.77	39.57	24.71
1994 - 95	11.44	1.50	554.00	188.98	-108.92	521.66	31.63	148.12	168.55
1995 - 96	160.89	1.06	653.07	271.67	-38.33	14.14	1.93	-7.42	132.13
1996 - 97	12.16	1.08	554.00	189.08	-185.75	126.27	-144.95	-68.14	60.47
1997 - 98	17.46	1.40	229.21	82.69	-221.30	258.38	-14.06	7.67	45.18
1998 - 99	20.44	1.43	981.89	334.59	-2698.94	-204.54	-261.15	-1054.88	-360.15
1999 - 00	20.47	1.45	325.59	115.84	-1838.04	300.45	-73.18	-536.92	-210.54
2000 - 01	24.47	2.16	505.17	177.27	-1626.06	540.65	-97.17	-394.19	-108.46
2001 - 02	21.10	2.46	184.22	69.26	-2042.19	50.94	-40.65	-677.30	-304.02
2002 - 03	23.15	-680.81	815.56	52.63	-1824.52	-86.59	-367.91	-759.67	-353.52
2003 - 04	785.07	-711.73	381.12	151.49	-1230.75	-164.99	12.95	-460.93	-154.72
2004 - 05	1519.75	-664.53	292.20	382.47	0.00	391.77	318.58	236.78	309.63
AVG	180.23	-135.72	367.71	137.41	-827.07	124.48	-5.32	-235.97	-49.28

Source: *Field Survey and Analyses of Sample Sugar Factories (Refer Nature and Source of Data Collection)*

APPENDIX - 22a: NET PROFIT (+) / LOSS (-) OF SELECTED SUGAR FACTORIES FROM 1990-91 TO 2004-05 (Rs in lakhs)

YEARS	PUBLIC SECTOR			PRIVATE SECTOR			OVERALL AVG (CO-OPERATIVE, PUBLIC AND PRIVATE)
	ABOVE AVG	BELOW AVG	AVG	ABOVE AVG	BELOW AVG	AVG	
	MY SUGAR	MYSORE		UGAR	DHAVANAGERE		
1990 – 91	161.26	0.00	80.63	74.82	34.79	54.80	65.33
1991 – 92	-26.18	329.47	151.64	86.10	28.03	57.07	85.54
1992 – 93	-225.42	0.00	-112.71	91.90	55.61	73.76	-45.29
1993 – 94	222.03	154.36	188.20	65.80	71.76	68.78	93.90
1994 – 95	359.32	610.25	484.79	320.51	43.50	182.01	278.45
1995 – 96	75.80	817.33	446.56	530.18	12.25	271.21	283.30
1996 – 97	-9.23	57.74	24.26	322.57	4.30	163.43	82.72
1997 – 98	-170.89	7.80	-81.55	354.42	42.00	198.21	53.95
1998 – 99	139.19	29.36	84.28	434.82	20.21	227.52	-16.12
1999 – 00	29.66	41.70	35.68	988.38	49.34	518.86	114.66
2000 – 01	-153.32	218.14	32.41	1616.73	79.38	848.06	257.33
2001 – 02	-67.54	-871.56	-469.55	51.00	6.62	28.81	-248.25
2002 – 03	-2588.37	-521.40	-1554.89	-120.09	13.24	-53.43	-653.94
2003 – 04	-1253.21	-244.54	-748.87	70.81	-314.29	-121.74	-341.78
2004 – 05	-20.10	138.41	59.15	53.83	-1057.85	-502.01	-44.41
AVG	-235.13	51.14	-92.00	329.45	-60.74	134.36	-2.31

Source: *Field Survey and Analyses of Sample Sugar Factories (Refer Nature and Source of Data Collection)*

APPENDIX -23: CURRENT RATIO FOR SAMPLE SUGAR FACTORIES FROM 1990-91 TO 2004-05

YEARS	CO-OPERATIVE SECTOR								
	ABOVE AVERAGE				BELOW AVERAGE				AVG
	HIRANYA-KESHI	MALA-PRABHA	DOOD-GHANGA	AVG	PANDAVA-PURA	GHATA-PRABHA	HEMA-VATHI	AVG	
1990 - 91	2.26	1.67	5.69	3.21		3.13	0.77	1.95	2.58
1991 - 92	2.08	1.69	4.94	2.90	4.64	3.17	1.27	3.03	2.96
1992 - 93	1.76	1.60	9.63	4.33	5.29	3.30	2.43	3.67	4.00
1993 - 94	1.74	1.22	2.59	1.85		2.77	2.08	2.43	2.14
1994 - 95	2.36	1.63	5.89	3.29	4.55	2.83	1.95	3.11	3.20
1995 - 96	2.83	1.75	1.02	1.87	7.24	2.41	2.22	3.96	2.91
1996 - 97	2.46	2.11	0.74	1.77	5.09	2.79	2.99	3.62	2.70
1997 - 98	2.62	1.66	0.49	1.59	0.47	8.52	2.92	3.97	2.78
1998 - 99	3.25	1.67	0.53	1.82	1.77	2.71	1.36	1.95	1.88
1999 - 00	5.81	2.05	0.55	2.80	2.10	3.24	2.18	2.50	2.65
2000 - 01	3.96	2.26	4.50	3.57	2.02	2.98	2.04	2.35	2.96
2001 - 02	4.72	5.09	5.85	5.22	3.70	4.40	2.24	3.45	4.33
2002 - 03	4.25	2.24	8.08	4.86	1.66	2.51	3.25	2.48	3.67
2003 - 04	9.11	1.71	9.04	6.62	1.93	6.66	2.35	3.65	5.13
2004 - 05	1.83	9.76	9.58	7.06		5.20	1.62	3.41	5.23
AVG	3.40	2.54	4.61	3.52	3.37	3.77	2.11	3.03	3.28

Source: *Ratio Analyses by Using Basic Data of Sample Sugar Factories*

APPENDIX -23a: CURRENT RATIO FOR SAMPLE SUGAR FACTORIES FROM 1990-91 TO 2004-05

YEARS	PUBLIC SECTOR			PRIVATE SECTOR			OVERALL AVG
	ABOVE AVG	*BELOW AVG*	*AVG*	*ABOVE AVG*	*BELOW AVG*	*AVG*	
	MY SUGAR	*MYSORE*		*UGAR*	*DHAVANAGERE*		
1990 - 91	1.46	2.97	1.46	1.89	1.30	1.60	1.88
1991 - 92	1.83	2.97	1.83	1.77	1.17	1.47	2.09
1992 - 93	1.91		1.91	1.71	1.20	1.46	2.46
1993 - 94	1.43	1.00	1.43	1.70	1.60	1.65	1.74
1994 - 95	1.60	2.77	1.60	1.92	3.66	2.79	2.53
1995 - 96	2.22	2.46	2.22	2.24	2.85	2.55	2.56
1996 - 97	3.59	3.05	3.59	1.64	4.64	3.14	3.14
1997 - 98	2.51	2.20	2.51	1.42	3.94	2.68	2.66
1998 - 99	2.22	1.87	2.22	3.02	2.89	2.95	2.35
1999 - 00	1.84	1.60	1.84	2.61	2.53	2.57	2.35
2000 - 01	1.87	1.70	1.87	2.54	2.89	2.72	2.51
2001 - 02	2.11	1.39	2.11	3.37	3.30	3.34	3.26
2002 - 03	1.54	0.14	1.54	2.92	3.91	3.42	2.87
2003 - 04	1.50	1.47	1.50	1.81	2.09	1.95	2.86
2004 - 05	0.12	1.39	0.12	2.52	2.79	2.65	2.67
AVG	1.85	1.93	1.85	2.21	2.72	2.46	2.53

Source: *Ratio Analyses by Using Basic Data of Sample Sugar Factories*

APPENDIX -24: ACID TEST RATIOS FOR SAMPLE SUGAR FACTORIES FROM 1990-91 TO 2004-05

YEARS	CO-OPERATIVE SECTOR								
	ABOVE AVERAGE				BELOW AVERAGE				AVG
	HIRANYA-KESHI	MALA-PRABHA	DOOD-GHANGA	AVG	PANDAVA-PURA	GHATA-PRABHA	HEMA-VATHI	AVG	
1990 - 91	2.26	1.67	5.69	3.21		3.13	0.77	1.95	2.58
1991 - 92	2.08	1.69	4.94	2.90	4.64	3.17	1.27	3.03	2.96
1992 - 93	1.76	1.60	9.63	4.33	5.29	3.30	2.43	3.67	4.00
1993 - 94	1.74	1.22	2.59	1.85		2.77	2.08	2.43	2.14
1994 - 95	2.36	1.63	5.89	3.29	4.55	2.83	1.95	3.11	3.20
1995 - 96	2.83	1.75	1.02	1.87	7.24	2.41	2.22	3.96	2.91
1996 - 97	2.46	2.11	0.74	1.77	5.09	2.79	2.99	3.62	2.70
1997 - 98	2.62	1.66	0.49	1.59	0.47	8.52	2.92	3.97	2.78
1998 - 99	3.25	1.67	0.53	1.82	1.77	2.71	1.36	1.95	1.88
1999 - 00	5.81	2.05	0.55	2.80	2.10	3.24	2.18	2.50	2.65
2000 - 01	3.96	2.26	4.50	3.57	2.02	2.98	2.04	2.35	2.96
2001 - 02	4.72	5.09	5.85	5.22	3.70	4.40	2.24	3.45	4.33
2002 - 03	4.25	2.24	8.08	4.86	1.66	2.51	3.25	2.48	3.67
2003 - 04	9.11	1.71	9.04	6.62	1.93	6.66	2.35	3.65	5.13
2004 - 05	1.83	9.76	9.58	7.06		5.20	1.62	3.41	5.23
AVG	3.40	2.54	4.61	3.52	3.37	3.77	2.11	3.03	3.28

Source: *Ratio Analyses by Using Basic Data of Sample Sugar Factories*

APPENDIX -24a ACID TEST RATIOS FOR SAMPLE SUGAR FACTORIES FROM 1990-91 TO 2004-05

YEARS	PUBLIC SECTOR			PRIVATE SECTOR			OVERALL AVG
	ABOVE AVG	BELOW AVG	AVG	ABOVE AVG	BELOW AVG	AVG	
	MY SUGAR	MYSORE		UGAR	DHAVANAGERE		
1990 - 91	1.46	2.97	2.22	1.89	1.30	1.60	2.13
1991 - 92	1.83	2.97	2.40	1.77	1.17	1.47	2.28
1992 - 93	1.91		1.91	1.71	1.20	1.46	2.46
1993 - 94	1.43	1.00	1.22	1.70	1.60	1.65	1.67
1994 - 95	1.60	2.77	2.19	1.92	3.66	2.79	2.72
1995 - 96	2.22	2.46	2.34	2.24	2.85	2.55	2.60
1996 - 97	3.59	3.05	3.32	1.64	4.64	3.14	3.05
1997 - 98	2.51	2.20	2.36	1.42	3.94	2.68	2.61
1998 - 99	2.22	1.87	2.04	3.02	2.89	2.95	2.29
1999 - 00	1.84	1.60	1.72	2.61	2.53	2.57	2.31
2000 - 01	1.87	1.70	1.78	2.54	2.89	2.72	2.49
2001 - 02	2.11	1.39	1.75	3.37	3.30	3.34	3.14
2002 - 03	1.54	0.14	0.84	2.92	3.91	3.42	2.64
2003 - 04	1.50	1.47	1.49	1.81	2.09	1.95	2.86
2004 - 05	0.12	1.39	0.76	2.52	2.79	2.65	2.88
AVG	1.85	1.93	1.89	2.21	2.72	2.46	2.54

Source: *Ratio Analyses by Using Basic Data of Sample Sugar Factories*

APPENDIX -25: LIQUID ASSETS TO TOTAL ASSETS RATIO FOR SAMPLE SUGAR FACTORIES FROM 1990-91 TO 2004-05

YEARS	CO-OPERATIVE SECTOR								
	ABOVE AVERAGE				*BELOW AVERAGE*				*AVG*
	HIRANYA-KESHI	*MALA-PRABHA*	*DOOD-GHANGA*	*AVG*	*PANDAVA-PURA*	*GHATA-PRABHA*	*HEMA-VATHI*	*AVG*	
1990 - 91	0.61	0.71	0.12	0.48		0.53	0.21	0.37	0.43
1991 - 92	0.60	0.76	0.12	0.49	0.50	0.55	0.42	0.49	0.49
1992 - 93	0.67	0.75	0.12	0.51	0.38	0.59	0.45	0.48	0.49
1993 - 94	0.63	0.78	0.21	0.54		0.54	0.50	0.52	0.53
1994 - 95	0.60	0.77	0.31	0.56	0.39	0.60	0.57	0.52	0.54
1995 - 96	0.64	0.78	0.08	0.50	0.53	0.61	0.59	0.58	0.54
1996 - 97	0.67	0.79	0.08	0.52	0.22	0.62	0.58	0.47	0.49
1997 - 98	0.67	0.71	0.06	0.48	0.05	0.58	0.47	0.36	0.42
1998 - 99	0.62	0.74	0.08	0.48	0.15	0.61	0.57	0.44	0.46
1999 - 00	0.97	0.76	0.09	0.61	0.19	0.65	0.62	0.49	0.55
2000 - 01	0.64	0.79	0.61	0.68	0.22	0.70	0.70	0.54	0.61
2001 - 02	0.68	0.85	0.61	0.71	0.17	0.72	0.66	0.51	0.61
2002 - 03	0.61	0.72	0.58	0.64	0.16	0.56	0.65	0.46	0.55
2003 - 04	0.55	0.61	0.48	0.55	0.08	0.68	0.65	0.47	0.51
2004 - 05	0.42	0.76	0.56	0.58		0.71	0.49	0.60	0.59
AVG	0.64	0.75	0.28	0.56	0.25	0.62	0.54	0.49	0.52

Source: *Ratio Analyses by Using Basic Data of Sample Sugar Factories*

APPENDIX -25a LIQUID ASSETS TO TOTAL ASSETS RATIO FOR SAMPLE SUGAR FACTORIES FROM 1990-91 TO 2004-05

YEARS	PUBLIC SECTOR			PRIVATE SECTOR			OVERALL AVG
	ABOVE AVG	BELOW AVG	AVG	ABOVE AVG	BELOW AVG	AVG	
	MY SUGAR	MYSORE		UGAR	DHAVANAGERE		
1990 - 91	1.22	1.00	1.11	0.70	0.89	0.80	0.78
1991 - 92	1.51	1.00	1.26	0.70	0.90	0.80	0.85
1992 - 93	1.42	1.00	1.21	0.89	0.93	0.91	0.87
1993 - 94	1.66	1.00	1.33	0.71	0.87	0.79	0.88
1994 - 95	1.69	1.00	1.34	0.81	0.93	0.87	0.92
1995 - 96	1.33	1.00	1.17	0.78	0.91	0.85	0.85
1996 - 97	1.11	1.00	1.06	0.69	0.92	0.80	0.78
1997 - 98	1.10	1.00	1.05	0.47	0.93	0.70	0.72
1998 - 99	1.34	1.00	1.17	0.62	0.78	0.70	0.78
1999 - 00	1.60	1.00	1.30	0.71	0.66	0.69	0.84
2000 - 01	1.45	1.00	1.22	0.61	0.60	0.61	0.81
2001 - 02	1.22	1.00	1.11	0.60	0.70	0.65	0.79
2002 - 03	0.81	1.00	0.91	0.66	0.66	0.66	0.71
2003 - 04	7.70	1.00	4.35	0.53	0.45	0.49	1.78
2004 - 05	1.89	1.00	1.45	0.64	0.52	0.58	0.87
AVG	1.80	1.00	1.40	0.67	0.78	0.73	0.88

Source: *Ratio Analyses by Using Basic Data of Sample Sugar Factories*

APPENDIX-26: INVENTORY TO NET WORKING CAPITAL RATIOS FOR SAMPLE SUGAR FACTORIES FROM 1990-91 TO 2004-05

YEARS	CO-OPERATIVE SECTOR								
	ABOVE AVERAGE				BELOW AVERAGE				AVG
	HIRANYA-KESHI	MALA-PRABHA	DOOD-GHANGA	AVG	PANDAVA-PURA	GHATA-PRABHA	HEMA-VATHI	AVG	
1990 - 91	1.27	1.79	5.03	2.70		1.20	-1.11	0.05	1.37
1991 - 92	1.42	1.86	5.69	2.99	1.27	1.17	0.56	1.00	2.00
1992 - 93	1.81	2.07	5.65	3.18	1.12	1.32	0.93	1.12	2.15
1993 - 94	1.62	4.34	4.06	3.34		1.33	0.68	1.01	2.17
1994 - 95	1.26	2.00	1.47	1.58	1.07	1.37	0.46	0.97	1.27
1995 - 96	1.17	1.74	226.36	76.42	1.05	1.65	0.52	1.07	38.75
1996 - 97	1.27	1.37	-13.40	-3.59	1.04	1.49	1.50	1.34	-1.12
1997 - 98	1.16	1.69	-6.74	-1.30	-0.60	1.02	2.06	0.82	-0.24
1998 - 99	1.06	1.82	-5.20	-0.77	2.08	1.40	3.30	2.26	0.74
1999 - 00	0.63	1.47	-6.81	-1.57	1.77	1.32	1.58	1.56	-0.01
2000 - 01	1.07	1.42	1.12	1.20	1.59	1.00	1.28	1.29	1.25
2001 - 02	1.07	0.65	1.00	0.91	1.19	0.75	1.80	1.24	1.08
2002 - 03	1.10	1.26	0.84	1.07	2.32	1.29	1.51	1.71	1.39
2003 - 04	0.93	1.43	0.86	1.07	1.82	1.08	1.69	1.53	1.30
2004 - 05	1.56	0.37	0.64	0.86		1.15	4.11	2.63	1.74
AVG	1.23	1.69	14.71	5.87	1.31	1.24	1.39	1.31	3.59

Source: *Ratio Analyses by Using Basic Data of Sample Sugar Factories*

APPENDIX -26a: INVENTORY TO NET WORKING CAPITAL RATIOS FOR SAMPLE SUGAR FACTORIES FROM 1990-91 TO 2004-05

YEARS	PUBLIC SECTOR			PRIVATE SECTOR			OVERALL AVG
	ABOVE AVG	BELOW AVG	AVG	ABOVE AVG	BELOW AVG	AVG	
	MY SUGAR	MYSORE		UGAR	DHAVANAGERE		
1990 - 91	1.23	1.22	1.23	1.71	3.14	2.43	1.68
1991 - 92	0.86	7.02	3.94	1.78	5.54	3.66	3.20
1992 - 93	0.86	36.04	18.45	1.72	4.27	2.99	7.86
1993 - 94	2.42	0.73	1.57	1.69	1.88	1.79	1.84
1994 - 95	0.26	1.08	0.67	1.25	1.10	1.18	1.04
1995 - 96	0.83	1.24	1.04	1.18	1.36	1.27	13.69
1996 - 97	0.92	1.27	1.09	1.30	1.16	1.23	0.40
1997 - 98	1.93	1.21	1.57	1.76	0.83	1.29	0.88
1998 - 99	0.57	2.38	1.47	1.05	1.02	1.03	1.08
1999 - 00	0.85	3.23	2.04	0.98	1.32	1.15	1.06
2000 - 01	1.61	2.97	2.29	1.15	1.20	1.17	1.57
2001 - 02	1.44	4.63	3.03	0.90	1.17	1.04	1.72
2002 - 03	1.08	1.47	1.27	0.10	0.73	0.41	1.02
2003 - 04	0.98	2.23	1.60	1.55	0.56	1.05	1.32
2004 - 05	-0.04	2.72	1.34	1.06	0.67	0.87	1.32
AVG	1.05	4.63	2.84	1.28	1.73	1.50	2.64

Source: *Ratio Analyses by Using Basic Data of Sample Sugar Factories*

APPENDIX-27: TOTAL LIABILITIES TO OWNED FUNDS RATIO FOR SAMPLE SUGAR FACTORIES FROM 1990-91 TO 2004-05

YEARS	CO-OPERATIVE SECTOR								AVG
	ABOVE AVERAGE				BELOW AVERAGE				
	HIRANYA-KESHI	MALA-PRABHA	DOOD-GHANGA	AVG	PANDAVA-PURA	GHATA-PRABHA	HEMA-VATHI	AVG	
1990 - 91	16.51	13.76	1.45	10.57		0.98	0.69	0.83	5.70
1991 - 92	19.42	21.69	2.53	14.55	2.22	1.25	1.04	1.50	8.02
1992 - 93	30.30	19.92	1.12	17.11	2.71	1.32	0.92	1.65	9.38
1993 - 94	30.18	25.93	1.72	19.28		1.23	0.77	1.00	10.14
1994 - 95	32.39	21.13	3.75	19.09	2.05	1.74	1.18	1.66	10.37
1995 - 96	39.82	19.31	5.69	21.61	3.23	2.01	1.30	2.18	11.89
1996 - 97	36.07	23.97	6.24	22.10	8.55	1.95	1.00	3.83	12.96
1997 - 98	32.22	18.23	5.67	18.71	12.34	1.40	0.64	4.79	11.75
1998 - 99	27.64	18.48	5.58	17.24	17.34	1.76	1.07	6.73	11.98
1999 - 00	33.01	21.84	8.04	20.96	22.12	1.61	1.31	8.35	14.65
2000 - 01	36.40	25.94	8.68	23.67	25.12	1.97	1.42	9.51	16.59
2001 - 02	42.50	12.70	5.97	20.39	23.27	1.91	1.54	8.91	14.65
2002 - 03	32.65	8.98	6.04	15.89	25.68	1.88	1.18	9.58	12.73
2003 - 04	14.16	6.85	8.43	9.82	23.11	1.62	1.38	8.70	9.26
2004 - 05	21.56	4.78	11.87	12.74		1.72	0.68	1.20	6.97
AVG	29.66	17.57	5.52	17.58	13.98	1.62	1.07	4.69	11.14

Source: *Ratio Analyses by Using Basic Data of Sample Sugar Factories*

APPENDIX -27a: TOTAL LIABILITIES TO OWNED FUNDS RATIO FOR SAMPLE SUGAR FACTORIES FROM 1990-91 TO 2004-05

YEARS	*PUBLIC SECTOR*			*PRIVATE SECTOR*			*OVERALL AVG*
	ABOVE AVG	*BELOW AVG*	*AVG*	*ABOVE AVG*	*BELOW AVG*	*AVG*	
	MY SUGAR	*MYSORE*		*UGAR*	*DHAVANAGERE*		
1990 - 91	2.06	0.92	1.49	10.07	5.62	7.84	5.01
1991 - 92	2.90	1.01	1.95	10.89	6.51	8.70	6.23
1992 - 93	4.42	0.82	2.62	11.81	6.63	9.22	7.07
1993 - 94	3.68	1.95	2.81	13.34	3.31	8.32	7.09
1994 - 95	3.38	1.64	2.51	9.73	7.06	8.39	7.09
1995 - 96	4.25	1.49	2.87	8.31	9.20	8.76	7.84
1996 - 97	4.83	1.35	3.09	6.00	9.41	7.71	7.92
1997 - 98	3.13	1.73	2.43	6.17	7.00	6.59	6.92
1998 - 99	4.13	1.89	3.01	6.86	10.34	8.60	7.86
1999 - 00	7.03	2.17	4.60	6.00	11.31	8.65	9.30
2000 - 01	7.94	2.03	4.98	4.08	6.55	5.31	8.96
2001 - 02	8.59	2.56	5.58	4.46	8.02	6.24	8.82
2002 - 03	9.27	2.55	5.91	5.68	3.73	4.70	7.78
2003 - 04	12.85	2.65	7.75	6.69	7.54	7.12	8.04
2004 - 05	16.76	2.55	9.66	5.19	4.75	4.97	7.20
AVG	6.35	1.82	4.08	7.69	7.13	7.41	7.54

Source: *Ratio Analyses by Using Basic Data of Sample Sugar Factories*

APPENDIX -28: FIXED ASSETS TO OWNED FUNDS RATIOS FOR SAMPLE SUGAR FACTORIES FROM 1990-91 TO 2004-05

YEARS	CO-OPERATIVE SECTOR								
	ABOVE AVERAGE				BELOW AVERAGE				AVG
	HIRANYA-KESHI	MALA-PRABHA	DOOD-GHANGA	AVG	PANDAVA-PURA	GHATA-PRABHA	HEMA-VATHI	AVG	
1990 - 91	12.90	6.49	6.29	8.56	0.00	0.64	1.65	0.77	4.66
1991 - 92	14.49	6.55	7.07	9.37	1.68	0.84	2.13	1.55	5.46
1992 - 93	15.92	7.51	3.45	8.96	1.82	0.87	2.22	1.64	5.30
1993 - 94	16.97	7.74	3.89	9.53	0.00	0.96	1.51	0.82	5.18
1994 - 95	19.58	6.73	5.25	10.52	1.64	0.92	2.17	1.58	6.05
1995 - 96	20.85	5.65	9.13	11.88	2.03	1.12	2.37	1.84	6.86
1996 - 97	17.52	6.41	10.03	11.32	7.18	1.10	1.85	3.38	7.35
1997 - 98	17.50	7.41	10.78	11.90	12.96	1.05	1.47	5.16	8.53
1998 - 99	19.57	6.66	10.92	12.38	12.17	0.96	1.75	4.96	8.67
1999 - 00	21.06	6.95	15.57	14.53	12.18	0.87	1.95	5.00	9.77
2000 - 01	22.86	6.99	7.48	12.44	12.19	0.91	2.46	5.19	8.82
2001 - 02	21.94	3.72	5.07	10.25	11.64	0.82	2.13	4.86	7.55
2002 - 03	21.94	3.50	6.28	10.58	1.83	1.32	2.11	1.75	6.16
2003 - 04	9.84	3.69	8.86	7.46	11.64	0.88	1.92	4.82	6.14
2004 - 05	9.72	4.07	10.25	8.01		1.02	1.41	1.22	4.62
AVG	17.51	6.00	8.02	10.51	6.36	0.95	1.94	2.97	6.74

Source: *Ratio Analyses by Using Basic Data of Sample Sugar Factories*

APPENDIX -28a: FIXED ASSETS TO OWNED FUNDS RATIOS FOR SAMPLE SUGAR FACTORIES FROM 1990-91 TO 2004-05

YEARS	*PUBLIC SECTOR*			*PRIVATE SECTOR*			*OVERALL AVG*
	ABOVE AVG	*BELOW AVG*	*AVG*	*ABOVE AVG*	*BELOW AVG*	*AVG*	
	MY SUGAR	*MYSORE*		*UGAR*	*DHAVANAGERE*		
1990 - 91	2.08	2.66	2.37	3.92	0.53	2.22	3.09
1991 - 92	2.32	2.24	2.28	4.28	0.61	2.45	3.40
1992 - 93	2.51	2.24	2.38	1.43	0.48	0.96	2.88
1993 - 94	2.90	1.58	2.24	5.10	0.45	2.78	3.40
1994 - 95	2.16	1.23	1.69	1.94	0.54	1.24	2.99
1995 - 96	1.99	1.15	1.57	1.97	0.85	1.41	3.28
1996 - 97	2.06	1.04	1.55	2.11	0.79	1.45	3.45
1997 - 98	2.17	0.95	1.56	3.67	0.56	2.12	4.07
1998 - 99	1.84	0.97	1.40	2.98	2.41	2.70	4.26
1999 - 00	2.21	0.82	1.51	2.05	4.11	3.08	4.79
2000 - 01	2.89	1.82	2.36	1.99	2.99	2.49	4.55
2001 - 02	3.53	2.10	2.81	2.23	2.74	2.49	4.28
2002 - 03	4.42	2.02	3.22	2.30	1.59	1.94	3.78
2003 - 04	55.70	2.16	28.93	2.94	4.63	3.79	12.95
2004 - 05	6.86	2.32	4.59	2.19	2.50	2.34	3.85
AVG	6.38	1.69	4.03	2.74	1.72	2.23	4.33

Source: *Ratio Analyses by Using Basic Data of Sample Sugar Factories*

APPENDIX 29: DEBT EQUITY RATIOS FOR SAMPLE SUGAR FACTORIES FROM 1990-91 TO 2004-05

YEARS	CO-OPERATIVE SECTOR								
	ABOVE AVERAGE				BELOW AVERAGE				AVG
	HIRANYA-KESHI	MALA-PRABHA	DOOD-GHANGA	AVG	PANDAVA-PURA	GHATA-PRABHA	HEMA-VATHI	AVG	
1990 - 91	6.71	4.04	1.29	4.01	0.00	0.75	0.23	0.33	2.17
1991 - 92	7.72	9.71	2.33	6.59	1.71	0.93	0.33	0.99	3.79
1992 - 93	10.32	5.70	1.08	5.70	2.24	0.94	0.51	1.23	3.46
1993 - 94	11.24	3.65	1.32	5.40	0.00	0.82	0.41	0.41	2.91
1994 - 95	18.59	7.57	3.34	9.83	1.58	1.25	0.55	1.13	5.48
1995 - 96	25.86	7.87	4.89	12.87	2.70	1.30	0.67	1.55	7.21
1996 - 97	20.65	12.39	4.99	12.67	8.00	1.31	0.64	3.32	8.00
1997 - 98	18.14	7.24	4.20	9.86	10.60	1.23	0.40	4.08	6.97
1998 - 99	17.35	7.34	3.72	9.47	15.20	1.21	0.34	5.59	7.53
1999 - 00	22.78	11.16	5.38	13.11	19.26	1.11	0.75	7.04	10.08
2000 - 01	25.72	14.26	6.08	15.35	21.13	1.24	0.58	7.65	11.50
2001 - 02	32.37	8.54	4.59	15.17	21.56	1.43	0.92	7.97	11.57
2002 - 03	24.25	5.02	4.98	11.42	21.67	1.20	0.76	7.87	9.65
2003 - 04	12.71	3.43	7.52	7.89	21.43	1.33	0.85	7.87	7.88
2004 - 05	17.68	3.48	10.53	10.57		1.24	0.25	0.74	5.65
AVG	18.14	7.43	4.42	9.99	10.51	1.15	0.54	3.85	6.92

Source: *Ratio Analyses by Using Basic Data of Sample Sugar Factories*

APPENDIX 29a: DEBT EQUITY RATIOS FOR SAMPLE SUGAR FACTORIES FROM 1990-91 TO 2004-05

YEARS	PUBLIC SECTOR			PRIVATE SECTOR			OVERALL AVG
	ABOVE AVG	*BELOW AVG*	*AVG*	*ABOVE AVG*	*BELOW AVG*	*AVG*	
	MY SUGAR	*MYSORE*		*UGAR*	*DHAVANAGERE*		
1990 - 91	0.66	0.74	0.70	5.20	2.26	3.73	2.20
1991 - 92	1.13	0.77	0.95	5.32	1.88	3.60	2.78
1992 - 93	2.11	0.82	1.47	5.33	1.67	3.50	2.81
1993 - 94	1.16	0.89	1.03	6.03	1.47	3.75	2.56
1994 - 95	1.13	1.05	1.09	5.33	5.19	5.26	3.94
1995 - 96	2.28	0.84	1.56	5.16	6.16	5.66	4.81
1996 - 97	3.45	0.87	2.16	3.18	7.48	5.33	5.16
1997 - 98	1.87	1.16	1.52	3.89	5.20	4.55	4.34
1998 - 99	2.20	1.11	1.65	5.25	7.35	6.30	5.16
1999 - 00	3.29	1.23	2.26	4.11	8.11	6.11	6.15
2000 - 01	4.03	1.25	2.64	2.87	4.97	3.92	6.02
2001 - 02	5.06	1.27	3.16	3.48	6.12	4.80	6.51
2002 - 03	5.72	1.33	3.53	4.12	2.93	3.53	5.57
2003 - 04	8.21	1.37	4.79	4.84	5.76	5.30	5.99
2004 - 05	8.67	1.17	4.92	3.65	3.77	3.71	4.76
AVG	3.40	1.06	2.23	4.52	4.69	4.60	4.58

Source: *Ratio Analyses by Using Basic Data of Sample Sugar Factories*

APPENDIX 30: INVENTORY TURNOVER RATIOS FOR SAMPLE SUGAR FACTORIES FROM 1990-91 TO 2004-05

YEARS	CO-OPERATIVE SECTOR								
	ABOVE AVERAGE				BELOW AVERAGE				AVG
	HIRANYA-KESHI	MALA-PRABHA	DOOD-GHANGA	AVG	PANDAVA-PURA	GHATA-PRABHA	HEMA-VATHI	AVG	
1990 - 91	1.83	17.84	1.96	7.21	1.02	1.67	0.75	1.15	4.18
1991 - 92	1.22	20.23	1.68	7.71	1.65	2.04	0.69	1.46	4.59
1992 - 93	1.27	24.00	1.45	8.91	0.00	1.57	0.53	0.70	4.80
1993 - 94	1.81	30.01	1.50	11.11	1.66	2.06	1.04	1.59	6.35
1994 - 95	1.35	27.84	1.27	10.15	0.19	1.45	0.66	0.77	5.46
1995 - 96	1.59	27.46	1.09	10.05	1.55	1.90	0.55	1.33	5.69
1996 - 97	1.20	19.78	1.28	7.42	1.30	1.78	0.79	1.29	4.36
1997 - 98	1.28	22.93	1.17	8.46	0.47	1.67	1.10	1.08	4.77
1998 - 99	1.56	20.72	1.39	7.89	1.31	1.36	0.62	1.09	4.49
1999 - 00	1.38	29.90	1.15	10.81	0.90	1.37	0.64	0.97	5.89
2000 - 01	1.26	23.13	0.97	8.45	0.76	1.29	0.54	0.86	4.66
2001 - 02	0.88	23.34	0.92	8.38	0.37	1.14	0.67	0.73	4.56
2002 - 03	1.10	29.25	0.85	10.40	0.77	1.63	0.59	1.00	5.70
2003 - 04	0.98	20.12	0.77	7.29		0.94	0.56	0.75	4.02
2004 - 05	1.81	14.52	12.28	9.54		1.06	0.95	1.00	5.27
AVG	1.37	23.40	1.98	8.92	0.92	1.53	0.71	1.05	4.98

Source: *Ratio Analyses by Using Basic Data of Sample Sugar Factories*

APPENDIX 30a: INVENTORY TURNOVER RATIOS FOR SAMPLE SUGAR FACTORIES FROM 1990-91 TO 2004-05

YEARS	PUBLIC SECTOR			PRIVATE SECTOR			OVERALL AVG
	ABOVE AVG	BELOW AVG	AVG	ABOVE AVG	BELOW AVG	AVG	
	MY SUGAR	MYSORE		UGAR	DHAVANAGERE		
1990 - 91	2.83	0.83	1.83	1.62	0.96	1.29	2.43
1991 - 92	2.71	0.60	1.66	1.49	0.99	1.24	2.49
1992 - 93	1.84	0.55	1.19	1.52	0.79	1.16	2.39
1993 - 94	3.80	0.99	2.39	1.62	1.54	1.58	3.44
1994 - 95	4.82	1.00	2.91	2.76	0.22	1.49	3.29
1995 - 96	2.90	0.87	1.88	2.25	0.93	1.59	3.05
1996 - 97	2.16	0.92	1.54	2.69	0.88	1.79	2.56
1997 - 98	2.92	1.40	2.16	3.90	1.59	2.74	3.22
1998 - 99	3.22	1.24	2.23	2.03	2.25	2.14	2.95
1999 - 00	1.87	1.04	1.45	2.04	0.85	1.45	2.93
2000 - 01	1.89	0.92	1.41	1.79	1.41	1.60	2.56
2001 - 02	1.85	1.28	1.57	1.92	0.80	1.36	2.49
2002 - 03	2.81	2.64	2.72	1.54	1.60	1.57	3.33
2003 - 04	1.56	1.52	1.54	1.78	1.38	1.58	2.38
2004 - 05	0.90	1.04	0.97	1.80	0.92	1.36	2.53
AVG	2.54	1.12	1.83	2.05	1.14	1.59	2.80

Source: *Ratio Analyses by Using Basic Data of Sample Sugar Factories*

APPENDIX 31: TOTAL ASSETS TURNOVER RATIOS FOR SAMPLE SUGAR FACTORIES FROM 1990-91 TO 2004-05

YEARS	CO-OPERATIVE SECTOR								
	ABOVE AVERAGE				BELOW AVERAGE				AVG
	HIRANYA-KESHI	MALA-PRABHA	DOOD-GHANGA	AVG	PANDAVA-PURA	GHATA-PRABHA	HEMA-VATHI	AVG	
1990 - 91	0.68	0.74	0.90	0.78		0.62	4.39	2.50	1.64
1991 - 92	0.66	0.76	0.85	0.76	0.79	0.89	13.57	5.08	2.92
1992 - 93	0.57	0.86	0.75	0.73	0.00	0.73	2.16	0.96	0.85
1993 - 94	0.93	1.02	0.75	0.90		0.85	5.93	3.39	2.14
1994 - 95	0.61	0.94	0.46	0.67	0.10	0.70	6.95	2.58	1.63
1995 - 96	0.70	0.84	0.39	0.64	0.70	0.83	1.53	1.02	0.83
1996 - 97	0.59	0.79	0.50	0.63	0.19	0.81	1.36	0.79	0.71
1997 - 98	0.68	1.05	0.46	0.73	0.14	0.94	1.88	0.99	0.86
1998 - 99	0.80	0.65	0.57	0.67	0.26	0.71	4.51	1.82	1.25
1999 - 00	0.64	0.72	0.46	0.61	0.20	0.77	1.68	0.88	0.75
2000 - 01	0.63	0.49	0.47	0.53	0.15	0.74	1.23	0.71	0.62
2001 - 02	0.46	0.35	0.48	0.43	0.06	0.72	1.03	0.60	0.52
2002 - 03	0.68	0.81	0.38	0.62	0.09	0.97	1.05	0.70	0.66
2003 - 04	0.52	0.66	0.28	0.48	0.05	0.51	0.92	0.49	0.49
2004 - 05	0.91	0.28	0.44	0.54		0.61	1.22	0.92	0.73
AVG	0.67	0.73	0.54	0.65	0.23	0.76	3.29	1.56	1.11

Source: *Ratio Analyses by Using Basic Data of Sample Sugar Factories*

APPENDIX 31a: TOTAL ASSETS TURNOVER RATIOS FOR SAMPLE SUGAR FACTORIES FROM 1990-91 TO 2004-05

YEARS	PUBLIC SECTOR			PRIVATE SECTOR			OVERALL AVG
	ABOVE AVG	BELOW AVG	AVG	ABOVE AVG	BELOW AVG	AVG	
	MY SUGAR	MYSORE		UGAR	DHAVANAGERE		
1990 - 91	2.66	0.00	1.33	0.96	0.60	0.78	1.25
1991 - 92	2.35	0.40	1.37	0.89	0.92	0.90	1.73
1992 - 93	1.62	0.54	1.08	1.07	0.80	0.93	0.95
1993 - 94	4.54	0.69	2.62	0.85	2.71	1.78	2.18
1994 - 95	4.50	0.50	2.50	1.63	0.14	0.88	1.67
1995 - 96	2.76	0.67	1.71	1.16	0.88	1.02	1.19
1996 - 97	1.96	0.76	1.36	1.48	1.02	1.25	1.10
1997 - 98	3.29	0.84	2.06	1.34	1.12	1.23	1.38
1998 - 99	2.59	0.56	1.57	0.71	0.64	0.68	1.17
1999 - 00	1.94	0.59	1.27	0.90	0.40	0.65	0.89
2000 - 01	2.37	0.60	1.48	0.83	0.74	0.79	0.96
2001 - 02	1.90	0.94	1.42	0.87	0.45	0.66	0.86
2002 - 03	1.71	1.14	1.43	0.62	1.02	0.82	0.97
2003 - 04	10.28	0.55	5.42	0.84	0.33	0.58	2.16
2004 - 05	6.73	0.57	3.65	0.80	0.18	0.49	1.62
AVG	3.41	0.62	2.02	1.00	0.80	0.90	1.34

Source: *Ratio Analyses by Using Basic Data of Sample Sugar Factories*

APPENDIX 32: WORKING CAPITAL TURN OVER RATIOS FOR SAMPLE SUGAR FACTORIES FROM 1990-91 TO 2004-05

YEARS	CO-OPERATIVE SECTOR								
	ABOVE AVERAGE				BELOW AVERAGE				AVG
	HIRANYA-KESHI	MALA-PRABHA	DOOD-GHANGA	AVG	PANDAVA-PURA	GHATA-PRABHA	HEMA-VATHI	AVG	
1990 - 91	2.00	2.60	8.86	4.49		1.73	-11.84	-5.05	-0.28
1991 - 92	2.12	2.45	8.77	4.45	2.01	2.37	7.63	4.01	4.23
1992 - 93	1.99	3.04	7.19	4.08	0.00	1.77	1.98	1.25	2.66
1993 - 94	3.45	7.33	5.74	5.50		2.45	4.04	3.24	4.37
1994 - 95	1.76	3.17	1.76	2.23	0.34	1.79	2.38	1.50	1.87
1995 - 96	1.69	2.51	207.72	70.64	1.55	2.34	1.69	1.86	36.25
1996 - 97	1.48	1.89	-16.96	-4.53	1.05	2.05	2.07	1.72	-1.40
1997 - 98	1.64	3.69	-7.03	-0.57	-2.48	1.83	3.59	0.98	0.21
1998 - 99	1.87	2.20	-7.66	-1.19	3.89	1.85	4.07	3.27	1.04
1999 - 00	0.80	1.85	-6.55	-1.30	2.02	1.71	1.90	1.88	0.29
2000 - 01	1.32	1.12	0.98	1.14	1.38	1.57	1.52	1.49	1.32
2001 - 02	0.86	0.51	0.94	0.77	0.53	1.30	1.84	1.22	1.00
2002 - 03	1.45	2.03	0.75	1.41	1.43	2.85	1.32	1.86	1.64
2003 - 04	1.05	2.58	0.66	1.43	1.38	0.88	1.50	1.25	1.34
2004 - 05	4.76	0.41	0.88	2.02		1.07	5.00	3.04	2.53
AVG	1.88	2.49	13.74	6.04	1.09	1.84	1.91	1.57	3.80

Source: *Ratio Analyses by Using Basic Data of Sample Sugar Factories*

APPENDIX 32a: WORKING CAPITAL TURN OVER RATIOS FOR SAMPLE SUGAR FACTORIES FROM 1990-91 TO 2004-05

YEARS	PUBLIC SECTOR			PRIVATE SECTOR			OVERALL AVG
	ABOVE AVG	*BELOW AVG*	*AVG*	*ABOVE AVG*	*BELOW AVG*	*AVG*	
	MY SUGAR	*MYSORE*		*UGAR*	*DHAVANAGERE*		
1990 - 91	6.91	2.42	4.66	2.91	2.91	2.91	2.43
1991 - 92	3.44	7.32	5.38	2.94	7.16	5.05	4.88
1992 - 93	2.39	36.03	19.21	2.91	5.10	4.00	8.63
1993 - 94	9.07	3.46	6.27	2.91	8.32	5.61	5.42
1994 - 95	7.08	1.65	4.37	4.19	0.20	2.20	2.81
1995 - 96	3.76	2.17	2.96	2.67	1.49	2.08	13.76
1996 - 97	2.44	2.57	2.51	5.50	1.41	3.45	1.52
1997 - 98	4.97	4.25	4.61	9.67	1.62	5.65	3.49
1998 - 99	3.51	3.43	3.47	1.72	1.25	1.48	2.00
1999 - 00	2.67	4.06	3.37	2.08	1.00	1.54	1.73
2000 - 01	3.52	3.88	3.70	2.25	1.88	2.06	2.36
2001 - 02	2.96	9.76	6.36	2.05	0.92	1.49	2.95
2002 - 03	6.01	30.82	18.42	1.42	2.06	1.74	7.26
2003 - 04	4.00	4.63	4.31	3.51	1.40	2.45	2.70
2004 - 05	-0.49	5.48	2.49	2.06	0.54	1.30	2.11
AVG	4.15	8.13	6.14	3.25	2.49	2.87	4.27

Source: *Ratio Analyses by Using Basic Data of Sample Sugar Factories*

APPENDIX 33: EFFICIENCY TO CAPITAL RATIOS FOR SAMPLE SUGAR FACTORIES FROM 1990-91 TO 2004-05

YEARS	CO-OPERATIVE SECTOR								
	ABOVE AVERAGE				BELOW AVERAGE				AVG
	HIRANYA-KESHI	MALA-PRABHA	DOOD-GHANGA	AVG	PANDAVA-PURA	GHATA-PRABHA	HEMA-VATHI	AVG	
1990 - 91	0.03	0.06	0.16	0.08	0.40	0.30	0.28	0.32	0.20
1991 - 92	0.03	0.05	0.15	0.07	0.24	0.19	0.22	0.22	0.15
1992 - 93	0.03	0.04	0.34	0.14		0.20	0.25	0.23	0.18
1993 - 94	0.02	0.03	0.27	0.10	0.20	0.20	0.25	0.22	0.16
1994 - 95	0.02	0.04	0.29	0.12	2.20	0.21	0.27	0.90	0.51
1995 - 96	0.02	0.05	0.26	0.11	0.18	0.14	0.26	0.19	0.15
1996 - 97	0.02	0.04	0.19	0.08	0.43	0.14	0.21	0.26	0.17
1997 - 98	0.02	0.04	0.20	0.08	0.42	0.15	0.19	0.26	0.17
1998 - 99	0.02	0.06	0.15	0.08	0.16	0.17	0.27	0.20	0.14
1999 - 00	0.02	0.05	0.13	0.06	0.15	0.15	0.29	0.20	0.13
2000 - 01	0.02	0.06	0.11	0.06	0.17	0.12	0.27	0.19	0.13
2001 - 02	0.02	0.12	0.17	0.10	0.38	0.13	0.25	0.25	0.18
2002 - 03	0.02	0.10	0.18	0.10	0.23	0.10	0.32	0.22	0.16
2003 - 04	0.07	0.16	0.21	0.15	0.40	0.19	0.36	0.32	0.23
2004 - 05	0.05	0.14	0.12	0.10		0.13	0.30	0.21	0.16
AVG	0.03	0.07	0.19	0.10	0.43	0.17	0.27	0.28	0.19

Source: *Ratio Analyses by Using Basic Data of Sample Sugar Factories*

APPENDIX 33a: EFFICIENCY TO CAPITAL RATIOS FOR SAMPLE SUGAR FACTORIES FROM 1990-91 TO 2004-05

YEARS	*PUBLIC SECTOR*			*PRIVATE SECTOR*			*OVERALL AVG*
	ABOVE AVG	*BELOW AVG*	*AVG*	*ABOVE AVG*	*BELOW AVG*	*AVG*	
	MY SUGAR	*MYSORE*		*UGAR*	*DHAVANAGERE*		
1990 - 91	0.09	1.05	0.57	0.03	0.30	0.16	0.31
1991 - 92	0.07	0.71	0.39	0.03	0.16	0.09	0.21
1992 - 93	0.08	0.60	0.34	0.03	0.18	0.10	0.21
1993 - 94	0.05	0.64	0.35	0.02	0.10	0.06	0.19
1994 - 95	0.04	0.92	0.48	0.02	1.50	0.76	0.58
1995 - 96	0.05	0.61	0.33	0.03	0.11	0.07	0.18
1996 - 97	0.05	0.52	0.29	0.02	0.09	0.06	0.17
1997 - 98	0.08	0.41	0.24	0.03	0.11	0.07	0.16
1998 - 99	0.10	0.59	0.35	0.04	0.13	0.09	0.19
1999 - 00	0.10	0.58	0.34	0.03	0.20	0.11	0.19
2000 - 01	0.06	0.56	0.31	0.03	0.13	0.08	0.17
2001 - 02	0.07	0.43	0.25	0.03	0.18	0.11	0.18
2002 - 03	0.07	0.37	0.22	0.03	0.18	0.11	0.16
2003 - 04	0.15	0.78	0.46	0.03	0.32	0.18	0.29
2004 - 05	0.98	0.78	0.88	0.03	1.05	0.54	0.52
AVG	0.14	0.64	0.39	0.03	0.32	0.17	0.25

Source: *Ratio Analyses by Using Basic Data of Sample Sugar Factories*

APPENDIX-34: GROSS RATIOS FOR SAMPLE SUGAR FACTORIES FROM 1990-91 TO 2004-05

YEARS	CO-OPERATIVE SECTOR								
	ABOVE AVERAGE				BELOW AVERAGE				AVG
	HIRANYA-KESHI	MALA-PRABHA	DOOD-GHANGA	AVG	PANDAVA-PURA	GHATA-PRABHA	HEMA-VATHI	AVG	
1990 - 91	99.65	232.45	64.97	132.35	240.99	124.56	116.30	160.62	146.49
1991 - 92	99.81	215.59	58.95	124.78	105.70	66.56	78.00	83.42	104.10
1992 - 93	99.86	247.18	58.86	135.30		158.57	100.80	129.68	132.49
1993 - 94	99.89	312.79	57.90	156.86	224.18	143.24	74.35	147.26	152.06
1994 - 95	99.92	249.51	53.77	134.40	117.94	272.43	101.82	164.06	149.23
1995 - 96	100.83	190.95	57.31	116.36	226.64	169.41	94.83	163.63	139.99
1996 - 97	99.94	229.91	51.09	126.98	169.15	126.36	85.96	127.16	127.07
1997 - 98	99.93	183.78	54.20	112.64	8987.99	102.85	5.27	3032.04	1572.34
1998 - 99	99.92	302.84	53.39	152.05	8473.84	320.48	108.60	2967.64	1559.85
1999 - 00	99.93	412.08	57.12	189.71	7951.52	109.22	95.91	2718.88	1454.29
2000 - 01	99.92	638.01	48.05	261.99	13022.56	132.06	215.33	4456.65	2359.32
2001 - 02	99.94	565.11	38.77	234.61	110.40	192.26	109.59	137.42	186.01
2002 - 03	99.93	-418.85	41.43	-92.50	672.06	292.31	2357.55	1107.31	507.40
2003 - 04	103.27	-317.47	30.05	-61.38	240.99	211.91	155.44	202.78	70.70
2004 - 05	455.06	-145.20	17.64	109.17		120.81	68.19	94.50	101.83
AVG	123.85	193.25	49.57	122.22	3118.77	169.53	251.19	1046.20	584.21

Source: *Ratio Analyses by Using Basic Data of Sample Sugar Factories*

APPENDIX -34a: GROSS RATIOS FOR SAMPLE SUGAR FACTORIES FROM 1990-91 TO 2004-05

YEARS	PUBLIC SECTOR			PRIVATE SECTOR			OVERALL AVG
	ABOVE AVG	BELOW AVG	AVG	ABOVE AVG	BELOW AVG	AVG	
	MY SUGAR	MYSORE		UGAR	DHAVANAGERE		
1990 - 91	93.63	87.93	90.78	108.54	95.05	101.79	113.02
1991 - 92	100.52	91.45	95.99	95.68	97.97	96.82	98.97
1992 - 93	104.41	100.42	102.42	96.47	95.95	96.21	110.37
1993 - 94	96.91	97.53	97.22	98.71	97.06	97.89	115.72
1994 - 95	97.31	91.63	94.47	99.71	97.46	98.59	114.10
1995 - 96	99.44	85.26	92.35	98.44	99.46	98.95	110.43
1996 - 97	104.70	99.19	101.95	98.26	99.81	99.04	109.35
1997 - 98	101.68	99.90	100.79	96.42	13.07	54.75	575.96
1998 - 99	99.07	99.63	99.35	97.78	98.62	98.20	585.80
1999 - 00	99.81	99.55	99.68	96.15	98.68	97.41	550.46
2000 - 01	100.86	9.80	55.33	93.33	98.45	95.89	836.85
2001 - 02	100.40	111.25	105.83	100.44	99.84	100.14	130.66
2002 - 03	117.45	106.12	111.78	99.35	99.82	99.58	239.59
2003 - 04	112.28	97.21	104.75	93.77	107.37	100.57	92.01
2004 - 05	144.10	1092.49	618.29	96.51	117.13	106.82	275.65
AVG	104.84	157.96	131.40	97.97	94.38	96.18	270.60

Source: *Ratio Analyses by Using Basic Data of Sample Sugar Factories*

APPENDIX–35: NET PROFITS TO OWNED FUNDS RATIOS FOR SAMPLE SUGAR FACTORIES FROM 1990-91 TO 2004-05

YEARS	CO-OPERATIVE SECTOR								
	ABOVE AVERAGE				BELOW AVERAGE				AVG
	HIRANYA-KESHI	MALA-PRABHA	DOOD-GHANGA	AVG	PANDAVA-PURA	GHATA-PRABHA	HEMA-VATHI	AVG	
1990 - 91	15.02	0.59	3.33	6.31	0.00	2.89	30.47	11.12	8.72
1991 - 92	9.35	1.17	0.72	3.75	-20.10	12.29	20.44	4.21	3.98
1992 - 93	8.35	1.81	1.24	3.80	-48.36	-35.57	7.65	-25.43	-10.81
1993 - 94	8.54	0.82	0.78	3.38	-67.70	25.48	0.68	-13.85	-5.23
1994 - 95	4.75	0.61	49.97	18.44	-23.83	32.44	2.34	3.65	11.05
1995 - 96	64.10	0.33	58.61	41.01	-8.38	0.90	0.13	-2.45	19.28
1996 - 97	3.72	0.33	49.71	17.92	-40.64	7.31	-8.57	-13.97	1.98
1997 - 98	4.80	0.43	18.90	8.05	-48.35	14.16	-0.84	-11.68	-1.82
1998 - 99	5.54	0.37	80.94	28.95	-589.65	-9.98	-14.62	-204.75	-87.90
1999 - 00	5.34	0.38	31.85	12.53	-401.55	14.26	-3.89	-130.40	-58.93
2000 - 01	6.26	0.56	49.40	18.74	-355.23	23.41	-5.14	-112.32	-46.79
2001 - 02	4.90	0.32	12.10	5.77	-425.43	1.93	-2.09	-141.86	-68.05
2002 - 03	5.24	-82.93	53.56	-8.04	-380.09	-3.64	-18.29	-134.01	-71.02
2003 - 04	79.73	-86.65	25.03	6.04	-256.39	-6.37	0.60	-87.38	-40.67
2004 - 05	150.91	-80.89	25.77	31.93		14.20	14.54	14.37	23.15
AVG	25.10	-16.18	30.79	13.24	-190.41	6.25	1.56	-56.32	-21.54

Source: *Ratio Analyses by Using Basic Data of Sample Sugar Factories*

APPENDIX -35a: NET PROFITS TO OWNED FUNDS RATIOS FOR SAMPLE SUGAR FACTORIES FROM 1990-91 TO 2004-05

YEARS	PUBLIC SECTOR			PRIVATE SECTOR			OVERALL AVG
	ABOVE AVG	*BELOW AVG*	*AVG*	*ABOVE AVG*	*BELOW AVG*	*AVG*	
	MY SUGAR	*MYSORE*		*UGAR*	*DHAVANAGERE*		
1990 - 91	17.81	0.00	8.90	12.72	15.22	13.97	10.53
1991 - 92	-3.04	10.17	3.57	14.49	11.46	12.98	6.84
1992 - 93	-27.67	0.00	-13.83	15.26	21.20	18.23	-2.14
1993 - 94	27.25	5.12	16.18	10.48	27.32	18.90	9.95
1994 - 95	29.34	22.22	25.78	32.23	16.41	24.32	20.38
1995 - 96	5.85	24.28	15.06	36.46	4.61	20.53	18.29
1996 - 97	-0.72	1.69	0.49	19.16	1.62	10.39	4.28
1997 - 98	-13.12	0.23	-6.44	18.40	10.65	14.52	2.09
1998 - 99	8.51	0.86	4.68	17.44	5.13	11.28	-23.98
1999 - 00	1.79	1.06	1.43	30.16	8.56	19.36	-12.72
2000 - 01	-10.32	5.40	-2.46	34.57	8.21	21.39	-9.29
2001 - 02	-4.60	-26.23	-15.41	1.22	0.65	0.94	-27.51
2002 - 03	-156.11	-16.43	-86.27	-2.96	0.76	-1.10	-52.80
2003 - 04	-88.88	-7.73	-48.31	1.92	-17.32	-7.70	-32.23
2004 - 05	-1.44	4.36	1.46	1.26	-32.09	-15.41	3.07
AVG	-14.36	1.67	-6.34	16.19	5.49	10.84	-5.68

Source: *Ratio Analyses by Using Basic Data of Sample Sugar Factories*

APPENDIX–36: NET PROFITS TO TOTAL ASSETS RATIOS FOR SAMPLE SUGAR FACTORIES FROM 1990-91 TO 2004-05

YEARS	CO-OPERATIVE SECTOR								
	ABOVE AVERAGE				BELOW AVERAGE				AVG
	HIRANYA-KESHI	MALA-PRABHA	DOOD-GHANGA	AVG	PANDAVA-PURA	GHATA-PRABHA	HEMA-VATHI	AVG	
1990 - 91	0.41	0.03	0.46	0.30		3.43	107.43	55.43	27.87
1991 - 92	0.23	0.04	0.09	0.12	-4.24	7.46	189.35	64.19	32.16
1992 - 93	0.16	0.06	0.32	0.18	-7.38	-22.73	14.04	-5.36	-2.59
1993 - 94	0.16	0.02	0.16	0.12		14.22	2.55	8.38	4.25
1994 - 95	0.09	0.02	6.53	2.21	-4.38	20.12	11.33	9.02	5.62
1995 - 96	1.04	0.01	5.89	2.31	-1.14	0.38	0.15	-0.20	1.05
1996 - 97	0.07	0.01	4.54	1.54	-3.22	3.12	-7.94	-2.68	-0.57
1997 - 98	0.09	0.02	1.64	0.58	-3.02	6.09	-0.97	0.70	0.64
1998 - 99	0.10	0.01	6.80	2.31	-23.62	-5.75	-61.03	-30.13	-13.91
1999 - 00	0.09	0.01	1.87	0.66	-12.76	7.37	-5.26	-3.55	-1.45
2000 - 01	0.09	0.02	2.58	0.90	-9.46	10.34	-4.77	-1.30	-0.20
2001 - 02	0.07	0.01	0.92	0.33	-11.31	0.92	-1.50	-3.96	-1.81
2002 - 03	0.09	-6.69	3.60	-1.00	-8.93	-1.25	-15.47	-8.55	-4.77
2003 - 04	3.31	-9.08	1.46	-1.44	-6.23	-4.49	0.52	-3.40	-2.42
2004 - 05	8.97	-4.83	1.12	1.75		6.59	13.29	9.94	5.85
AVG	1.00	-1.36	2.53	0.72	-7.97	3.06	16.11	5.90	3.31

Source: *Ratio Analyses by Using Basic Data of Sample Sugar Factories*

APPENDIX –36a: NET PROFITS TO TOTAL ASSETS RATIOS FOR SAMPLE SUGAR FACTORIES FROM 1990-91 TO 2004-05

YEARS	PUBLIC SECTOR			PRIVATE SECTOR			OVERALL AVG
	ABOVE AVG	BELOW AVG	AVG	ABOVE AVG	BELOW AVG	AVG	
	MY SUGAR	MYSORE		UGAR	DHAVANAGERE		
1990 - 91	10.70	0.00	5.35	0.97	3.11	2.04	11.75
1991 - 92	-1.43	5.73	2.15	1.03	1.90	1.47	11.93
1992 - 93	-8.88	0.00	-4.44	1.22	3.29	2.26	-1.59
1993 - 94	12.52	2.70	7.61	0.60	8.02	4.31	5.39
1994 - 95	13.75	10.79	12.27	3.11	2.22	2.67	6.85
1995 - 96	1.78	13.13	7.46	4.03	0.48	2.26	3.59
1996 - 97	-0.16	0.90	0.37	2.84	0.17	1.50	0.43
1997 - 98	-4.57	0.11	-2.23	2.66	1.39	2.03	0.15
1998 - 99	2.66	0.38	1.52	2.22	0.46	1.34	-3.68
1999 - 00	0.42	0.47	0.45	4.31	0.70	2.51	0.50
2000 - 01	-2.05	2.39	0.17	6.81	1.09	3.95	1.31
2001 - 02	-0.75	-11.52	-6.14	0.22	0.07	0.15	-2.60
2002 - 03	-23.24	-6.98	-15.11	-0.43	0.16	-0.14	-6.67
2003 - 04	-98.29	-3.23	-50.76	0.31	-2.07	-0.88	-18.02
2004 - 05	-2.76	1.88	-0.44	0.21	-6.12	-2.96	0.82
AVG	-6.69	1.12	-2.78	2.01	0.99	1.50	0.68

Source: *Ratio Analyses by Using Basic Data of Sample Sugar Factories*

APPENDIX -37: NET PROFITS TO TOTAL SALES RATIOS FOR SAMPLE SUGAR FACTORIES FROM 1990-91 TO 2004-05

YEARS	CO-OPERATIVE SECTOR								
	ABOVE AVERAGE				BELOW AVERAGE				AVG
	HIRANYA-KESHI	MALA-PRABHA	DOOD-GHANGA	AVG	PANDAVA-PURA	GHATA-PRABHA	HEMA-VATHI	AVG	
1990 - 91	0.61	0.04	0.51	0.39	0.00	3.43	24.49	9.31	4.85
1991 - 92	0.35	0.06	0.11	0.17	-5.36	7.46	13.96	5.35	2.76
1992 - 93	0.28	0.07	0.42	0.26		-22.73	6.50	-8.11	-3.93
1993 - 94	0.18	0.02	0.21	0.14	-13.30	14.22	0.43	0.45	0.29
1994 - 95	0.14	0.02	14.25	4.81	-42.05	20.12	1.63	-6.76	-0.98
1995 - 96	1.49	0.02	15.11	5.54	-1.62	0.38	0.10	-0.38	2.58
1996 - 97	0.11	0.01	9.12	3.08	-17.23	3.12	-5.83	-6.65	-1.78
1997 - 98	0.13	0.02	3.56	1.23	-21.11	6.09	-0.52	-5.18	-1.97
1998 - 99	0.13	0.02	12.01	4.05	-92.28	-5.75	-13.54	-37.19	-16.57
1999 - 00	0.14	0.02	4.10	1.42	-63.56	7.37	-3.12	-19.77	-9.18
2000 - 01	0.15	0.03	5.53	1.91	-63.01	10.34	-3.86	-18.85	-8.47
2001 - 02	0.15	0.04	1.93	0.70	-175.04	0.92	-1.46	-58.53	-28.91
2002 - 03	0.13	-8.29	9.51	0.45	-99.87	-1.25	-14.67	-38.60	-19.07
2003 - 04	6.43	-13.81	5.21	-0.72	-118.81	-4.49	0.56	-40.91	-20.82
2004 - 05	9.84	-17.14	2.56	-1.58		6.59	10.87	8.73	3.57
AVG	1.35	-2.59	5.61	1.46	-54.87	3.06	1.04	-14.47	-6.51

Source: *Ratio Analyses by Using Basic Data of Sample Sugar Factories*

APPENDIX -37a: NET PROFITS TO TOTAL SALES RATIOS FOR SAMPLE SUGAR FACTORIES FROM 1990-91 TO 2004-05

YEARS	PUBLIC SECTOR			PRIVATE SECTOR			OVERALL AVG
	ABOVE AVG	BELOW AVG	AVG	ABOVE AVG	BELOW AVG	AVG	
	MY SUGAR	MYSORE		UGAR	DHAVANAGERE		
1990 - 91	4.02	0.00	2.01	1.01	5.15	3.08	3.31
1991 - 92	-0.61	14.40	6.90	1.16	2.08	1.62	3.76
1992 - 93	-5.48	0.00	-2.74	1.15	4.10	2.62	-1.35
1993 - 94	2.76	3.89	3.32	0.71	2.96	1.83	1.82
1994 - 95	3.05	21.55	12.30	1.91	16.16	9.04	6.79
1995 - 96	0.65	19.69	10.17	3.49	0.55	2.02	4.92
1996 - 97	-0.08	1.18	0.55	1.93	0.16	1.04	-0.06
1997 - 98	-1.39	0.13	-0.63	1.98	1.24	1.61	-0.33
1998 - 99	1.03	0.68	0.85	3.13	0.73	1.93	-4.60
1999 - 00	0.21	0.80	0.51	4.77	1.75	3.26	-1.80
2000 - 01	-0.87	3.98	1.56	8.20	1.47	4.83	-0.69
2001 - 02	-0.40	-12.27	-6.33	0.25	0.16	0.21	-11.68
2002 - 03	-13.56	-6.14	-9.85	-0.70	0.16	-0.27	-9.73
2003 - 04	-9.56	-5.85	-7.70	0.36	-6.35	-2.99	-10.50
2004 - 05	-0.41	3.31	1.45	0.26	-33.54	-16.64	-3.87
AVG	-1.38	3.02	0.82	1.97	-0.21	0.88	-1.60

Source: *Ratio Analyses by Using Basic Data of Sample Sugar Factories*

APPENDIX -38: NET WORTH FOR SAMPLE SUGAR FACTORIES FROM 1990-91 TO 2004-05

YEARS	CO-OPERATIVE SECTOR								
	ABOVE AVERAGE				BELOW AVERAGE				AVG
	HIRANYA-KESHI	MALA-PRABHA	DOOD-GHANGA	AVG	PANDAVA-PURA	GHATA-PRABHA	HEMA-VATHI	AVG	
1990 - 91	3991.22	1754.85	2449.77	2731.95	0.00	617.32	856.05	491.12	1611.54
1991 - 92	4361.86	1015.30	2368.84	2582.00	1151.10	771.50	1046.04	989.55	1785.77
1992 - 93	4837.52	2050.50	3068.85	3318.96	1752.52	1091.05	1337.42	1393.66	2356.31
1993 - 94	4987.28	1777.74	3569.38	3444.80	0.00	1337.19	959.29	765.49	2105.15
1994 - 95	5234.38	1891.19	4329.76	3818.44	1548.31	930.80	1335.82	1271.64	2545.04
1995 - 96	5529.43	2061.52	4746.20	4112.38	1878.17	1284.03	1637.03	1599.74	2856.06
1996 - 97	6712.58	2226.54	5252.48	4730.53	1859.37	1612.70	1441.44	1637.84	3184.18
1997 - 98	8402.08	2411.46	7069.54	5961.03	1683.13	1975.55	1409.93	1689.54	3825.28
1998 - 99	9721.82	2609.98	7661.51	6664.44	3486.77	1402.37	1217.97	2035.70	4350.07
1999 - 00	10768.58	2684.38	9200.82	7551.26	4278.17	1885.67	1200.71	2454.85	5003.06
2000 - 01	11556.81	2876.85	10733.86	8389.18	5682.30	2554.62	1967.10	3401.34	5895.26
2001 - 02	12027.79	9393.28	10947.42	10789.50	6882.98	2622.46	1147.81	3551.08	7170.29
2002 - 03	11260.98	2796.65	13441.87	9166.50	8098.65	2734.33	1870.62	4234.53	6700.52
2003 - 04	9746.01	2207.75	13204.41	8386.05	8665.32	3029.31	1159.71	4284.78	6335.42
2004 - 05	-4769.05	9832.77	12697.52	5920.41	0.00	4956.89	1609.90	2188.93	4054.67
AVG	6957.95	3172.72	7382.82	5837.83	3131.12	1920.39	1346.46	2132.65	3985.24

Source: *Field Survey and Analyses of Sample Sugar Factories (Refer Nature and Source of Data Collection, Page No. 57, 58 & 59)*

APPENDIX–38a: NET WORTH FOR SAMPLE SUGAR FACTORIES FROM 1990-91 TO 2004-05

YEARS	*PUBLIC SECTOR*			*PRIVATE SECTOR*			*OVERALL AVG*
	ABOVE AVG	*BELOW AVG*	*AVG*	*ABOVE AVG*	*BELOW AVG*	*AVG*	
	MY SUGAR	*MYSORE*		*UGAR*	*DHAVANAGERE*		
1990 - 91	-357.88	610531.82	305086.97	1789.07	-164.39	812.34	102503.62
1991 - 92	-658.21	2471.85	906.82	1913.30	-120.42	896.44	1196.34
1992 - 93	-1065.79	2821.10	877.66	409.45	-47.05	181.20	1138.39
1993 - 94	-1224.31	-145.45	-684.88	2628.65	25.12	1326.88	915.72
1994 - 95	-1522.56	1149.05	-186.76	625.12	88.97	357.05	905.11
1995 - 96	-1254.50	1196.20	-29.15	1048.43	83.81	566.12	1131.01
1996 - 97	-481.71	1820.64	669.46	1232.08	88.10	660.09	1504.58
1997 - 98	-333.07	1496.31	581.62	1422.21	258.32	840.26	1749.05
1998 - 99	-1526.46	1262.00	-132.23	2460.60	278.79	1369.69	1862.51
1999 - 00	-4539.23	281.56	-2128.83	3243.14	510.58	1876.86	1583.70
2000 - 01	-4314.36	922.31	-1696.02	4652.76	956.21	2804.48	2334.57
2001 - 02	-3648.82	-952.37	-2300.60	4555.33	1006.42	2780.87	2550.19
2002 - 03	-4227.52	-634.47	-2430.99	4705.56	1730.59	3218.07	2495.87
2003 - 04	-16835.64	-807.20	-8821.42	-1450.95	1494.88	21.97	-821.35
2004 - 05	-22741.47	-739.49	-11740.48	3775.72	1607.03	2691.38	-1664.81
AVG	-4315.44	41378.26	18531.41	2200.70	519.80	1360.25	7958.97

Source: *Field Survey and Analyses of Sample Sugar Factories (Refer Nature and Source of Data Collection, Page No. 57, 58 & 59)*

APPENDIX -39: NET CAPITAL RATIOS FOR SAMPLE SUGAR FACTORIES FROM 1990-91 TO 2004-05

YEARS	CO-OPERATIVE SECTOR								
	ABOVE AVERAGE				BELOW AVERAGE				AVG
	HIRANYA-KESHI	MALA-PRABHA	DOOD-GHANGA	AVG	PANDAVA-PURA	GHATA-PRABHA	HEMA-VATHI	AVG	
1990 - 91	2.20	1.65	4.97	2.94		1.39	0.41	0.90	1.919
1991 - 92	2.07	1.24	3.18	2.16	2.14	1.48	0.10	1.24	1.702
1992 - 93	1.74	1.52	3.47	2.25	2.42	1.62	0.59	1.54	1.894
1993 - 94	1.73	1.34	2.87	1.98		1.71	0.35	1.03	1.506
1994 - 95	1.67	1.36	2.04	1.69	2.65	1.33	0.17	1.39	1.539
1995 - 96	1.55	1.33	1.75	1.54	2.27	1.41	0.66	1.44	1.495
1996 - 97	1.57	1.29	1.76	1.54	1.48	1.48	1.08	1.35	1.441
1997 - 98	1.72	1.41	2.03	1.72	1.30	1.77	1.35	1.48	1.597
1998 - 99	1.95	1.37	2.13	1.82	1.44	1.39	0.22	1.02	1.418
1999 - 00	1.85	1.32	2.12	1.76	1.42	1.55	0.56	1.18	1.472
2000 - 01	1.81	1.29	2.21	1.77	1.49	1.56	0.76	1.27	1.520
2001 - 02	1.66	1.96	2.20	1.94	1.62	1.52	0.90	1.35	1.643
2002 - 03	1.78	1.38	2.46	1.87	1.66	1.61	1.00	1.42	1.649
2003 - 04	1.70	1.39	2.03	1.71	1.78	1.72	0.84	1.45	1.578
2004 - 05	0.78	3.50	1.94	2.08		2.05	1.62	1.83	1.954
AVG	1.72	1.56	2.48	1.92	1.80	1.57	0.71	1.33	1.62

Source: *Ratio Analyses by Using Basic Data of Sample Sugar Factories*

APPENDIX -39a: NET CAPITAL RATIOS FOR SAMPLE SUGAR FACTORIES FROM 1990-91 TO 2004-05

YEARS	PUBLIC SECTOR			PRIVATE SECTOR			OVERALL AVG
	ABOVE AVG	BELOW AVG	AVG	ABOVE AVG	BELOW AVG	AVG	
	MY SUGAR	MYSORE		UGAR	DHAVANAGERE		
1990 - 91	0.81	199.83	100.32	1.30	0.87	1.09	34.44
1991 - 92	0.74	1.75	1.25	1.30	0.92	1.11	1.35
1992 - 93	0.70	2.23	1.47	1.06	0.97	1.02	1.46
1993 - 94	0.59	0.98	0.78	1.31	1.03	1.17	1.15
1994 - 95	0.63	1.26	0.94	1.06	1.05	1.06	1.18
1995 - 96	0.77	1.24	1.00	1.09	1.03	1.06	1.19
1996 - 97	0.92	1.40	1.16	1.12	1.04	1.08	1.23
1997 - 98	0.92	1.25	1.09	1.12	1.09	1.11	1.26
1998 - 99	0.77	1.19	0.98	1.14	1.07	1.11	1.17
1999 - 00	0.61	1.03	0.82	1.16	1.08	1.12	1.14
2000 - 01	0.63	1.11	0.87	1.24	1.15	1.20	1.20
2001 - 02	0.71	0.89	0.80	1.24	1.12	1.18	1.21
2002 - 03	0.72	0.92	0.82	1.20	1.27	1.24	1.24
2003 - 04	0.07	0.90	0.49	0.94	1.11	1.03	1.03
2004 - 05	0.03	0.91	0.47	1.17	1.10	1.14	1.19
AVG	0.64	14.46	7.55	1.17	1.06	1.11	3.43

Source: *Ratio Analyses by Using Basic Data of Sample Sugar Factories*

APPENDIX–40: EQUITY TO CAPITAL RATIOS FOR SAMPLE SUGAR FACTORIES FROM 1990-91 TO 2004-05

YEARS	CO-OPERATIVE SECTOR								
	ABOVE AVERAGE				BELOW AVERAGE				AVG
	HIRANYA-KESHI	MALA-PRABHA	DOOD-GHANGA	AVG	PANDAVA-PURA	GHATA-PRABHA	HEMA-VATHI	AVG	
1990 - 91	0.98	1.00	0.60	0.86	1.00	1.00	1.00	1.00	0.93
1991 - 92	0.98	1.00	0.66	0.88	1.00	1.00	1.00	1.00	0.94
1992 - 93	0.98	1.00	0.26	0.75	1.00	1.00	1.00	1.00	0.87
1993 - 94	0.98	1.00	0.44	0.81	1.00	1.00	1.00	1.00	0.90
1994 - 95	0.98	1.00	0.51	0.83	1.00	1.00	1.00	1.00	0.91
1995 - 96	0.98	1.00	0.53	0.83	1.00	1.00	1.00	1.00	0.92
1996 - 97	0.99	1.00	0.53	0.84	1.00	1.00	1.00	1.00	0.92
1997 - 98	0.97	1.00	0.49	0.82	1.00	1.00	1.00	1.00	0.91
1998 - 99	0.97	1.00	0.49	0.82	1.00	1.00	1.00	1.00	0.91
1999 - 00	0.97	1.00	0.59	0.85	1.00	1.00	1.00	1.00	0.93
2000 - 01	0.97	1.00	0.51	0.83	1.00	1.00	1.00	1.00	0.91
2001 - 02	0.97	1.00	0.67	0.88	1.00	1.00	0.99	1.00	0.94
2002 - 03	0.97	1.00	0.75	0.91	1.00	1.00	1.00	1.00	0.95
2003 - 04	0.99	1.00	0.77	0.92	1.00	1.00	1.00	1.00	0.96
2004 - 05	1.00	1.00	1.00	1.00		1.00	1.00	1.00	1.00
AVG	0.98	1.00	0.59	0.85	1.00	1.00	1.00	1.00	0.93

Source: *Ratio Analyses by Using Basic Data of Sample Sugar Factories*

APPENDIX -40a: EQUITY TO CAPITAL RATIOS FOR SAMPLE SUGAR FACTORIES FROM 1990-91 TO 2004-05

YEARS	PUBLIC SECTOR			PRIVATE SECTOR			OVERALL AVG
	ABOVE AVG	BELOW AVG	AVG	ABOVE AVG	BELOW AVG	AVG	
	MY SUGAR	MYSORE		UGAR	DHAVANAGERE		
1990 - 91	1.00	0.94	0.97	1.00	1.00	1.00	0.97
1991 - 92	1.00	1.00	1.00	1.00	1.00	1.00	0.98
1992 - 93	1.00	1.00	1.00	1.00	1.00	1.00	0.96
1993 - 94	1.00	1.00	1.00	1.00	1.00	1.00	0.97
1994 - 95	1.00	1.00	1.00	1.00	1.00	1.00	0.97
1995 - 96	1.00	1.00	1.00	1.00	1.00	1.00	0.97
1996 - 97	1.00	1.00	1.00	1.00	1.00	1.00	0.97
1997 - 98	1.00	1.00	1.00	1.00	1.00	1.00	0.97
1998 - 99	1.00	1.00	1.00	1.00	1.00	1.00	0.97
1999 - 00	1.00	1.00	1.00	1.00	1.00	1.00	0.98
2000 - 01	1.00	1.00	1.00	1.00	1.00	1.00	0.97
2001 - 02	1.00	1.00	1.00	1.00	1.00	1.00	0.98
2002 - 03	1.00	1.00	1.00	1.00	1.00	1.00	0.98
2003 - 04	1.00	1.00	1.00	1.00	1.00	1.00	0.99
2004 - 05	1.00	1.00	1.00	1.00	1.00	1.00	1.00
AVG	1.00	1.00	1.00	1.00	1.00	1.00	0.98

Source: *Ratio Analyses by Using Basic Data of Sample Sugar Factories*

APPENDIX–41: FIXED ASSETS TO TOTAL ASSETS FOR SAMPLE SUGAR FACTORIES FROM 1990-91 TO 2004-05

YEARS	CO-OPERATIVE SECTOR								
	ABOVE AVERAGE				BELOW AVERAGE				AVG
	HIRANYA-KESHI	MALA-PRABHA	DOOD-GHANGA	AVG	PANDAVA-PURA	GHATA-PRABHA	HEMA-VATHI	AVG	
1990 - 91	0.36	0.29	0.88	0.51		0.47	5.82	3.15	1.83
1991 - 92	0.36	0.24	0.88	0.49	0.35	0.45	19.73	6.84	3.67
1992 - 93	0.30	0.25	0.88	0.48	0.28	0.41	4.08	1.59	1.03
1993 - 94	0.32	0.22	0.79	0.44		0.46	5.68	3.07	1.76
1994 - 95	0.36	0.23	0.69	0.43	0.30	0.40	10.51	3.74	2.08
1995 - 96	0.34	0.22	0.92	0.49	0.28	0.39	2.78	1.15	0.82
1996 - 97	0.31	0.21	0.92	0.48	0.57	0.38	1.71	0.89	0.68
1997 - 98	0.32	0.29	0.94	0.51	0.81	0.42	1.71	0.98	0.75
1998 - 99	0.36	0.26	0.92	0.51	0.49	0.39	7.32	2.73	1.62
1999 - 00	0.34	0.24	0.91	0.50	0.39	0.35	2.63	1.12	0.81
2000 - 01	0.35	0.21	0.39	0.32	0.32	0.30	2.28	0.97	0.64
2001 - 02	0.31	0.15	0.39	0.28	0.31	0.28	1.53	0.71	0.49
2002 - 03	0.38	0.28	0.42	0.36	0.04	0.44	1.78	0.75	0.56
2003 - 04	0.41	0.39	0.52	0.44	0.28	0.32	1.65	0.75	0.59
2004 - 05	0.58	0.24	0.44	0.42		0.29	1.29	0.79	0.61
AVG	0.36	0.25	0.72	0.44	0.37	0.38	4.70	1.95	1.20

Source: *Ratio Analyses by Using Basic Data of Sample Sugar Factories*

APPENDIX-41a: FIXED ASSETS TO TOTAL ASSETS FOR SAMPLE SUGAR FACTORIES FROM 1990-91 TO 2004-05

YEARS	PUBLIC SECTOR			PRIVATE SECTOR			OVERALL AVG
	ABOVE AVG	BELOW AVG	AVG	ABOVE AVG	BELOW AVG	AVG	
	MY SUGAR	MYSORE		UGAR	DHAVANAGERE		
1990 - 91	1.25	0.01	0.63	0.30	0.11	0.20	0.89
1991 - 92	1.09	1.26	1.17	0.30	0.10	0.20	1.68
1992 - 93	0.81	1.23	1.02	0.11	0.07	0.09	0.71
1993 - 94	1.33	0.83	1.08	0.29	0.13	0.21	1.02
1994 - 95	1.01	0.60	0.80	0.19	0.07	0.13	1.01
1995 - 96	0.61	0.62	0.61	0.22	0.09	0.15	0.53
1996 - 97	0.46	0.55	0.51	0.31	0.08	0.20	0.46
1997 - 98	0.76	0.44	0.60	0.53	0.07	0.30	0.55
1998 - 99	0.57	0.43	0.50	0.38	0.22	0.30	0.81
1999 - 00	0.51	0.37	0.44	0.29	0.34	0.31	0.52
2000 - 01	0.57	0.81	0.69	0.39	0.40	0.39	0.58
2001 - 02	0.58	0.92	0.75	0.40	0.30	0.35	0.53
2002 - 03	0.66	0.86	0.76	0.34	0.34	0.34	0.55
2003 - 04	61.60	0.90	31.25	0.47	0.55	0.51	10.78
2004 - 05	13.21	1.00	7.11	0.36	0.48	0.42	2.71
AVG	5.67	0.72	3.19	0.33	0.22	0.27	1.56

Source: *Ratio Analyses by Using Basic Data of Sample Sugar Factories*

APPENDIX-42: FIXED ASSETS TO NET WORTH RATIOS FOR SAMPLE SUGAR FACTORIES FROM 1990-91 TO 2004-05

YEARS	CO-OPERATIVE SECTOR								
	ABOVE AVERAGE				BELOW AVERAGE				AVG
	HIRANYA-KESHI	MALA-PRABHA	DOOD-GHANGA	AVG	PANDAVA-PURA	GHATA-PRABHA	HEMA-VATHI	AVG	
1990 - 91	0.65	0.73	1.10	0.83		1.70	1.71	1.71	1.27
1991 - 92	0.69	1.27	1.28	1.08	0.67	1.39	1.95	1.34	1.21
1992 - 93	0.71	0.72	1.24	0.89	0.47	1.07	1.71	1.08	0.99
1993 - 94	0.77	0.87	1.21	0.95		1.10	2.04	1.57	1.26
1994 - 95	0.90	0.88	1.35	1.04	0.48	1.60	2.20	1.43	1.23
1995 - 96	0.95	0.88	2.14	1.32	0.49	1.37	2.22	1.36	1.34
1996 - 97	0.85	0.93	2.13	1.30	1.77	1.18	2.17	1.70	1.50
1997 - 98	0.76	0.99	1.85	1.20	3.52	0.97	1.76	2.08	1.64
1998 - 99	0.74	0.98	1.73	1.15	1.60	1.40	2.57	1.86	1.50
1999 - 00	0.75	0.99	1.73	1.16	1.30	0.98	3.05	1.78	1.47
2000 - 01	0.77	0.94	0.71	0.81	0.98	0.82	2.36	1.39	1.10
2001 - 02	0.79	0.31	0.71	0.60	0.81	0.82	3.61	1.75	1.17
2002 - 03	0.86	1.03	0.71	0.87	0.11	1.15	2.27	1.17	1.02
2003 - 04	0.99	1.37	1.02	1.13	0.65	0.76	3.55	1.65	1.39
2004 - 05	-2.05	0.34	0.92	-0.27		0.57	1.92	1.25	0.49
AVG	0.61	0.88	1.32	0.94	1.07	1.12	2.34	1.54	1.24

Source: *Ratio Analyses by Using Basic Data of Sample Sugar Factories*

APPENDIX-42a: FIXED ASSETS TO NET WORTH RATIOS FOR SAMPLE SUGAR FACTORIES FROM 1990-91 TO 2004-05

YEARS	PUBLIC SECTOR			PRIVATE SECTOR			OVERALL AVG
	ABOVE AVG	BELOW AVG	AVG	ABOVE AVG	BELOW AVG	AVG	
	MY SUGAR	MYSORE		UGAR	DHAVANAGERE		
1990 - 91	5.26	0.01	-2.62	1.29	-0.73	0.28	-0.36
1991 - 92	-3.03	2.93	-0.05	1.33	-1.25	0.04	0.40
1992 - 93	-1.92	2.22	0.15	2.10	-2.68	-0.29	0.28
1993 - 94	-1.93	-32.77	-17.35	1.22	4.75	2.98	-4.37
1994 - 95	-1.73	2.93	0.60	3.08	1.60	2.34	1.39
1995 - 96	-2.06	3.22	0.58	2.74	2.70	2.72	1.55
1996 - 97	-5.46	1.95	-1.76	2.88	2.38	2.63	0.79
1997 - 98	-8.49	2.16	-3.17	4.97	0.86	2.92	0.46
1998 - 99	-1.97	2.62	0.33	3.02	3.41	3.22	1.68
1999 - 00	-0.80	11.46	5.33	2.07	4.64	3.35	3.38
2000 - 01	-1.00	7.97	3.49	2.00	3.02	2.51	2.37
2001 - 02	-1.42	-7.31	-4.37	2.04	2.76	2.40	-0.26
2002 - 03	-1.74	-10.08	-5.91	1.98	1.59	1.78	-1.03
2003 - 04	-4.66	-8.48	-6.57	-7.46	5.62	-0.92	-2.03
2004 - 05	-0.42	-9.97	-5.20	2.48	5.12	3.80	-0.30
AVG	-2.79	-2.07	-2.43	1.71	2.25	1.98	0.26

Source: *Ratio Analyses by Using Basic Data of Sample Sugar Factories*

APPENDIX-43: FIXED ASSETS TO TOTAL SALES RATIOS FOR SAMPLE SUGAR FACTORIES FROM 1990-91 TO 2004-05

YEARS	CO-OPERATIVE SECTOR								
	ABOVE AVERAGE				*BELOW AVERAGE*				*AVG*
	HIRANYA-KESHI	*MALA-PRABHA*	*DOOD-GHANGA*	*AVG*	*PANDAVA-PURA*	*GHATA-PRABHA*	*HEMA-VATHI*	*AVG*	
1990 - 91	0.52	0.39	0.97	0.63	0.00	0.77	1.33	0.70	0.66
1991 - 92	0.54	0.32	1.03	0.63	0.45	0.51	1.45	0.80	0.72
1992 - 93	0.53	0.29	1.17	0.66		0.56	1.89	1.22	0.94
1993 - 94	0.35	0.22	1.05	0.54	0.00	0.54	0.96	0.50	0.52
1994 - 95	0.59	0.25	1.50	0.78	2.89	0.57	1.51	1.66	1.22
1995 - 96	0.48	0.26	2.35	1.03	0.39	0.48	1.81	0.89	0.96
1996 - 97	0.52	0.26	1.84	0.88	3.04	0.47	1.26	1.59	1.23
1997 - 98	0.47	0.28	2.03	0.92	5.66	0.45	0.91	2.34	1.63
1998 - 99	0.45	0.40	1.62	0.82	1.90	0.55	1.62	1.36	1.09
1999 - 00	0.54	0.33	2.00	0.96	1.93	0.45	1.57	1.32	1.14
2000 - 01	0.55	0.42	0.84	0.60	2.16	0.40	1.85	1.47	1.04
2001 - 02	0.68	0.43	0.81	0.64	4.79	0.39	1.49	2.22	1.43
2002 - 03	0.55	0.35	1.12	0.67	0.48	0.45	1.69	0.87	0.77
2003 - 04	0.79	0.59	1.85	1.08	5.40	0.62	1.79	2.60	1.84
2004 - 05	0.63	0.86	1.02	0.84		0.47	1.06	0.77	0.80
AVG	0.55	0.38	1.41	0.78	2.24	0.51	1.48	1.35	1.07

Source: *Ratio Analyses by Using Basic Data of Sample Sugar Factories*

APPENDIX -43a: FIXED ASSETS TO TOTAL SALES RATIOS FOR SAMPLE SUGAR FACTORIES FROM 1990-91 TO 2004-05

YEARS	PUBLIC SECTOR			PRIVATE SECTOR			OVERALL AVG
	ABOVE AVG	BELOW AVG	AVG	ABOVE AVG	BELOW AVG	AVG	
	MY SUGAR	MYSORE		UGAR	DHAVANAGERE		
1990 - 91	0.47	5.35	2.91	0.31	0.18	0.25	1.27
1991 - 92	0.46	3.17	1.82	0.34	0.11	0.23	0.92
1992 - 93	0.50	2.26	1.38	0.11	0.09	0.10	0.81
1993 - 94	0.29	1.20	0.75	0.34	0.05	0.20	0.49
1994 - 95	0.22	1.19	0.71	0.11	0.53	0.32	0.75
1995 - 96	0.22	0.93	0.57	0.19	0.10	0.15	0.56
1996 - 97	0.24	0.73	0.48	0.21	0.08	0.15	0.62
1997 - 98	0.23	0.52	0.38	0.40	0.07	0.23	0.75
1998 - 99	0.22	0.76	0.49	0.54	0.34	0.44	0.67
1999 - 00	0.26	0.62	0.44	0.32	0.84	0.58	0.72
2000 - 01	0.24	1.34	0.79	0.47	0.54	0.50	0.78
2001 - 02	0.30	0.98	0.64	0.46	0.68	0.57	0.88
2002 - 03	0.38	0.75	0.57	0.54	0.33	0.44	0.59
2003 - 04	5.99	1.64	3.81	0.56	1.70	1.13	2.26
2004 - 05	1.96	1.76	1.86	0.45	2.61	1.53	1.40
AVG	0.80	1.55	1.17	0.36	0.55	0.45	0.90

Source: *Ratio Analyses by Using Basic Data of Sample Sugar Factories*

Example

APPENDIX-44: SUGAR PRODUCTION OF SELECTED SUGAR COMPANIES OF THE YEAR 1990 - 91

	X	Mean	X-Mean	$(X\text{-Mean})^2$
Hiranyakeshi	793602	500747.8	292854.2	85763582457.64
Malaprabha	668690	500747.8	167942.2	28204582540.84
Doodghanga	483854	500747.8	-16893.8	285400478.44
Pandavapura	274870	500747.8	-225877.8	51020780532.84
Ghataprabha	300030	500747.8	-200717.8	40287635236.84
Hemavathi	141250	500747.8	-359497.8	129238668204.84
My Sugar	602840	500747.8	102092.2	10422817300.84
Mysore	380405	500747.8	-120342.8	14482389511.84
Ugar	1176130	500747.8	675382.2	456141116076.84
Davangere	185807	500747.8	-314940.8	99187707504.64
Mean	500747.80		" $(X\text{-Mean})^2$	915034679845.60

$$\frac{\Sigma (X\text{-Mean})^2}{N-1} = 101670519982.84$$

STANDARD ERROR = $\frac{\Sigma (X\text{-Mean})^2}{N-1}$ = 318858.1503

APPENDIX–45: PHYSICAL AND FINANCIAL INDICATORS SELECTED FROM THE RESEARCH WORK TO DETERMINE THE STANDARD ERROR

Physical Indicators

1. Sugar Production
2. Quantity of Sugarcane Crushed
3. Recovery Percentage
4. Capacity Utilisation

Financial Indicators

1. Share Capital
2. Net Profit
3. Barrowings
4. Current Liabilities
5. Level of reserve funds
6. Total expenses
7. Gross income
8. Total assets position
9. Total sales
10. Fixed assets position

Procedure:

Taken Sugar Production as one Physical Indicator or Variable (X) of Selected Companies.

1. Calculated the Mean Value of 1990-91.
2. Deducted Mean Value from Individual Sample Value i.e. (X-Mean)
3. Square the (X-Mean) Value
4. Obtained the Sum of Square of (X-Mean)
5. Divide the step 5 by (N-1) Where N is Number of Samples.
6. Taking Square Root of Step 6 obtains Standard Error.

APPENDIX - 46: CRITICAL VALUES OF STUDENT'S t- DISTRIBUTION

	LEVEL OF SIGNIFICANT FOR ONE-TAILED TEST						
d.f.	*.25*	*.10*	*.05*	*.025*	*.01*	*.005*	*.0005*
	LEVEL OF SIGNIFICANT FOR TWO-TAILED TEST						
v	*.50*	*.20*	*.10*	*.05*	*.02*	*.01*	*.001*
1	1.000	3.078	6.314	12.706	31.821	63.657	636.19
2	.816	1.886	2.920	4.303	6.965	9.925	31.599
3	.765	1.638	2.353	3.182	4.541	5.841	12.924
4	.741	1.533	2.132	2.776	3.747	4.604	8.610
5	.727	1.476	2.015	2.571	3.365	4.032	6.869
6	.718	1.440	1.943	2.447	3.143	3.707	5.959
7	.711	1.415	1.895	2.365	2.998	3.499	5.408
8	.706	1.397	1.860	2.306	2.896	3.355	5.041
9	.703	1.383	1.833	2.262	2.821	3.250	4.781
10	.700	1.372	1.812	2.228	2.764	3.169	4.587
11	.697	1.363	1.796	2.201	2.718	3.106	4.437
12	.695	1.356	1.782	2.179	2.681	3.055	4.318
13	.694	1.350	1.771	2.160	2.650	3.012	4.221
14	.692	1.345	1.761	2.145	2.624	2.977	4.140
15	.691	1.341	1.753	2.131	2.602	2.947	4.073

APPENDIX - 46: *Contd...*

16	.690	1.337	1.746	2.120	2.583	2.921	4.015
17	.689	1.333	1.740	2.110	2.567	2.898	3.965
18	.688	1.330	1.734	2.101	2.552	2.878	3.922
19	.688	1.328	1.729	2.093	2.539	2.861	3.883
20	.687	1.325	1.725	2.086	2.528	2.845	3.850
21	.686	1.323	1.721	2.080	2.518	2.831	3.819
22	.686	1.321	1.717	2.074	2.508	2.819	3.792
23	.685	1.319	1.714	2.069	2.500	2.807	3.768
24	.865	1.318	1.711	2.064	2.492	2.797	3.745
25	.684	1.316	1.708	2.060	2.485	2.787	3.725
26	.684	1.315	1.706	2.056	2.479	2.779	3.707
27	.684	1.314	1.703	2.052	2.473	2.771	3.690
28	.683	1.313	1.701	2.048	2.467	2.763	3.674
29	.683	1.311	1.699	2.045	2.462	2.756	3.659
30	.683	1.310	1.697	2.042	2.457	2.750	3.646
40	.681	1.303	1.684	2.021	2.423	2.704	3.551
60	.679	1.296	1.671	2.000	2.390	2.660	3.460
120	.677	1.289	1.658	1.980	2.358	2.617	3.373
∞	.674	1.282	1.645	1.960	2.326	2.576	3.291

INDEX

A

B

C

I

J

K

L

M

N

O

P

Q

R

S

T

U

V

W